MAX BROCKMAN

FUTURE SCIENCE

Max Brockman is the vice president of Brockman, Inc., a literary agency, and the editor of *What's Next? Dispatches on the Future of Science*. He also works with the Edge Foundation, Inc., a non-profit organization that publishes the Edge newsletter (www.edge.org). He lives in New York City.

FUTURE SCIENCE

FUTURE SCIENCE
ESSAYS FROM THE CUTTING EDGE

EDITED AND
WITH A PREFACE BY

MAX BROCKMAN

OXFORD
UNIVERSITY PRESS

OXFORD
UNIVERSITY PRESS

Great Clarendon Street, Oxford OX2 6DP

Oxford University Press is a department of the University of Oxford.
It furthers the University's objective of excellence in research, scholarship,
and education by publishing worldwide in

Oxford New York

Auckland Cape Town Dar es Salaam Hong Kong Karachi
Kuala Lumpur Madrid Melbourne Mexico City Nairobi
New Delhi Shanghai Taipei Toronto

With offices in

Argentina Austria Brazil Chile Czech Republic France Greece
Guatemala Hungary Italy Japan Poland Portugal Singapore
South Korea Switzerland Thailand Turkey Ukraine Vietnam

Oxford is a registered trade mark of Oxford University Press
in the UK and in certain other countries

Published in the United States
by Oxford University Press Inc., New York

British Library Cataloguing in Publication Data
Data available

Library of Congress Cataloging-in-Publication Data
Brockman, Max.
Future science : essays from the cutting edge / Max Brockman.
p. cm. —(Vintage original)
ISBN 978-0-19-969935-3 (pbk.)
1. Science—Forecasting. 2. Technological forecasting. I. Title.
Q171.B8726 2011
500—dc22
2011003006

Typeset by SPI Publisher Services, Pondicherry, India
Printed in Great Britain
on acid-free paper by
Clays Ltd, St Ives plc

ISBN 978-0-19-969935-3

1 3 5 7 9 10 8 6 4 2

To Jennie

CONTENTS

PREFACE

Academia, with its somewhat old-fashioned structure and rules, can appear quite a strange place when observed from the outside. Frequently, through my work as a literary agent, I've noticed that if you're an academic who writes about your work for a general audience, you're thought by some of your colleagues to be wasting your time and perhaps endangering your academic career. For younger scientists (i.e., those without tenure), this is almost universally true.

There are some good reasons for this peer pressure, the most obvious being that getting published in academic journals is an essential step on the very difficult road to tenure. However, one unfortunate result is that those of us outside academia are blocked from looking in on the research being done by this next generation of scientists, some of whom will go on to become leading doers and communicators of science.

This opacity was the impetus for the first essay collection in this series, *What's Next?: Dispatches on the Future of Science*. Essays seemed to be an ideal and appropriate way for representatives of this group of scientists to communicate their ideas. The title of the new collection is different, but the organization is the same. *Future Science* features essays from nineteen young scientists from a variety of fields, writing about what they're working on and what excites them the most. To come up with the list of contributors, I fielded recommendations from top scientists on the rising stars in their various disciplines.

Among those you will hear from in *Future Science* are:

- Kevin P. Hand, a planetary scientist and astrobiologist at the Jet Propulsion Laboratory in Pasadena, California, on the possibilities for life elsewhere in the solar system (and the universe)

- Felix Warneken, who heads the Social Cognitive Development Group at Harvard's Laboratory for Developmental Studies, on investigating the evolutionary roots of human altruism in his studies of young children and Ugandan chimpanzees

- William McEwan, a virologist and postdoctoral researcher at the MRC Laboratory of Molecular Biology, Cambridge, U.K., who probes the biology of antiviral immunity by designing his own viruses

- Anthony Aguirre, a physicist and cosmologist at the University of California, Santa Cruz, who maintains that infinity has been brought into the domain of testable physical science

- Daniela Kaufer and Darlene Francis of the University of California, Berkeley, neurobiologists who have taken a transdisciplinary approach to the study of the effects of stress on mind and body

- Jon Kleinberg, a professor of computer science at Cornell University, who is working on ways to extract significance from the enormous data sets we are building in the Internet age.

Working on *Future Science* has been an extremely rewarding experience, and I look forward to putting together the next collection in this series. These passionate young scientists, by giving us a glimpse of the work they're doing today, are in a sense providing a window into the world to come.

Max Brockman
New York
August 2011

FUTURE SCIENCE

KEVIN P. HAND,
a planetary scientist and astrobiologist at the Jet Propulsion Laboratory in Pasadena, California, received a BA in physics from Dartmouth College (1998), a master's in mechanical engineering from Stanford University (2002), and a PhD in geological and environmental sciences from Stanford (2007). His research focuses on the origin, evolution, and distribution of life in the solar system and involves both numerical modeling and experiments on the physics and chemistry of icy moons in the outer system, with an emphasis on Jupiter's Europa.

ON THE COMING AGE OF OCEAN EXPLORATION

KEVIN P. HAND

On a clear spring day in 2020, I hope to be staring at a rocket on a launchpad and listening to a countdown. The passenger atop that rocket will be a robotic spacecraft that, among other things, will represent one of the most advanced and extraordinary technological achievements ever created by humans. This passenger will endure a harsh eight-year journey en route to its final destination—in orbit around Europa, the sixth moon of Jupiter. Its goal will be nothing less than to explore the habitability of an ocean that may contain two to three times the volume of all the liquid water on Earth. Europa's ocean, like our own, has existed for much of the history of the Solar System. It's an ocean that could answer one of the oldest questions humans have ever asked: Is there life beyond Earth?

This mission marks the beginning of a new age in exploration. For much of human history, we have been explorers of continents—examiners of rocks and regions ripe for habitation, the culmination being the Heroic Age of Antarctic exploration and the capstone being our flags and footprints on the surface of the Moon. But in the decades and centuries to come, exploration—both human and robotic—will increasingly focus on the ocean depths, of both our own ocean and the subsurface oceans believed to exist on at least five moons of the outer Solar System: Jupiter's Europa, Ganymede, and Callisto and Saturn's Titan and Enceladus. The total volume of liquid water on those worlds is estimated to be more than a hundred times the volume of liquid water on Earth.

The motivation for the coming age of ocean exploration will be to better understand the delicate balance of Earth's habitability while simultaneously expanding our understanding of habitability—and possibly inhabitants—on distant worlds in our Solar System and beyond. It is an endeavor that will benefit greatly from both private entrepreneurs and inventors and large-scale initiatives by NASA and the National Science Foundation. NASA envisions autonomous vehicles that will melt through the surface ice and explore those distant oceans with a very high degree of artificial intelligence. Such missions are many decades away from the launchpad, but their technological precursors—some of which are funded by the NSF and others by entrepreneurs—are currently exploring Earth's ocean, discovering bizarre life-forms deep beneath the surface.

But beyond the prospect of liquid water, what makes ocean worlds like Europa compelling places for astrobiology? Despite considerable evidence to the contrary, Earth was not a particularly good place for life to arise. The main ingredients for life as we know it are a lot easier to find farther out in the Solar System. The reason goes back to the earliest days of planet formation: With the Sun at the center, the primordial disk of gas and dust differentiated as a result of heat, gravity, and solar wind. Heavy materials condensed and accreted into the planets of the inner Solar System, and lighter materials eventually "froze out" in the cold outer regions. Water, carbon dioxide, methane, ammonia, and nitrogen gas are all light, volatile compounds that were largely baked out of the inner Solar System and condensed into the gas giant planets and their moons. Those compounds are the raw materials of life.

The fact that Earth has a large ocean has long been a mystery to planetary scientists. Given Earth's location and formation history, it should be a considerably drier, less volatile-rich world. The generally agreed-upon solution to this puzzle is that after the epoch of planet formation, comets impacting Earth delivered water and other volatiles here. Indeed, it may be that for each glass of water you drink, roughly half comes from comets. Some of the evidence for this hypothesis was uncovered by studies of hydrogen isotopes in water—isotopes that serve as fingerprints identifying the water's original source. Much of our scientific interest in comets stems from a desire to better understand the role those objects played in making Earth a habitable world.

So the ice-covered ocean worlds of the outer Solar System likely harbor liquid water and the raw materials for life. Much of the water in the outer Solar System is ice, but some is liquid. On Europa, for example, we have good evidence for a global liquid-water ocean some 100 kilometers deep, lying beneath an ice shell perhaps no more than 20 kilometers thick. That's a tremendous amount of water, and if we've learned anything about life on Earth, it's that where you find liquid water, you generally find life.

How can such oceans exist? Why hasn't all of the water turned to ice on those cold, tiny worlds? The first thing to appreciate is the wonderful little fact that ice floats. If ice didn't float, those ocean worlds would likely have frozen over. But since ice does float, and since it serves as a great thermal blanket, any heat generated in the interior of those worlds can go into maintaining liquid-water oceans. On Europa, some of the internal heating comes, as it does on Earth, from the radioactive decay of heavy elements in the rocky mantle beneath the ocean. A larger source of energy, though, may come from the tidal tug and pull Europa experiences as it orbits Jupiter. Radioactive decay and tidal energy in various combinations likely serve as the dominant internal heating mechanisms for many of the outer Solar System satellites.

Understanding the evidence for the existence of their oceans is a little more complicated. We can't, of course, see the oceans in images taken from orbit because of the icy shells that blanket the surface. To be sure, the myriad ice fractures and disrupted terrain on the surfaces of some of the moons evoke comparisons to icebergs and broken ice sheets in our own polar regions. But such cracks could have been created many millions of years ago. How do we know those oceans persist today? For Europa and Callisto, some of the strongest evidence comes from an unusual source: measurements from the *Galileo* spacecraft of changes in the magnetic fields around these moons. The data show that neither has a strong internal magnetic field, like that of Earth, but rather that their fields are induced by the constantly changing magnetic field of Jupiter as it rotates on its axis once every ten hours.

It's the same physics that makes metal detectors at airports work. As you walk through the doorway of a metal detector, you are actually walking through a rapidly changing magnetic field. If you have a conducting material in your pocket, electric currents will be induced, and those currents will give rise to an induced magnetic field. Detectors in the doorway sense the induced field, and the alarm goes off. Well, on *Galileo,* the alarm basically went off as it flew by Europa and Callisto. The observation of induced magnetic fields on these moons indicates the presence of a near-surface conducting layer. It was known (from other lines of evidence) that the upper few hundred kilometers of those moons was either liquid or frozen water; the magnetic-field observations imply salty, subsurface liquid-water oceans, since ice is not nearly conductive enough to explain the induced field. Interestingly, conducting iron cores deep within the interiors of the moons cannot explain the data; the conducting region must be close to the surface.

While some detailed physics is required to discern the subsurface oceans on Europa and Callisto, sometimes you can just look at what's happening on the surface of such a world. This is the case for Enceladus. Imagery and other data from the *Cassini* spacecraft have revealed plumes of salt- and organic-rich water jetting into space. At this point, we don't have a complete story for what is happening on Enceladus, but evidence is mounting that the water of the plumes comes from a subsurface south polar sea.

We now have good reason to believe that most of the liquid water in the Solar System lies beneath the surfaces of those icy worlds, from tiny Enceladus (504 kilometers in diameter) to grand Ganymede (larger than Mercury). Those oceans may represent the bulk of habitable real estate in our Solar System—and, for that matter, ice-covered oceans may constitute the bulk of habitable real estate throughout the Universe. Some have argued that worlds like the Earth are rare; if our Solar System is any guide, ice-covered ocean worlds might be more abundant abodes of life.

But how could such an ecosystem thrive? How can life exist in an ocean covered with ice, cut off from the heat and light of the Sun? To answer that question, we return to the depths of our own ocean.

For much of the history of human exploration, Earth's ocean filled the role that outer space now plays. The oceanic world was the realm through which the human imagination soared; the public had a seemingly insatiable appetite for reports of new seafaring adventures and discoveries of bizarre and fantastic creatures and hitherto unknown civilizations. The abysses, the deepest depths of the ocean, were unknown and unknowable. It was a region devoid of the Sun's light and heat and therefore thought to be devoid of life.

But in the mid- to late 1800s, that understanding began to change. Expeditions like that of HMS *Challenger* managed to troll the depths with nets and dredge up thousands of never-before-seen animals. In the early decades of the twentieth century, the invention of the manned submersible enabled us to explore the subsurface with our own eyes. In the decades that followed, deep-sea exploration advanced in fits and starts, often prompted less by scientific interest than by a desire for military dominance of the seas. In early 1960, the United States laid claim to that dominance by touching down with the bathyscape *Trieste* at a depth of 35,797 feet in the Mariana Trench, the deepest known part of our ocean. Fifty years out from that achievement, although we have made only incremental progress in our ability to explore the depths, our scientific understanding of the ocean and seafloor has improved dramatically.

Between 1998 and 2008, the U.S. fleet of manned submersibles, robotically operated vehicles (ROVs), and autonomous underwater vehicles (AUVs) embarked on 231 expeditions—a marked increase in the average annual number of such expeditions over those from 1960 through 1997. If we make the rather generous assumption that each of those 231 expeditions provided a hundred hours of deep-sea exploration time, we find that the total amounts to almost three years. Though that may sound like a lot of time, it's not when you consider the vast unknown expanse of the seafloor. For comparison, the Lewis and Clark Expedition in 1804 took nearly three years to chart a path across the American West and unveiled countless vistas and monumental landscapes but hardly completed the process of exploring the region west of the Mississippi. Now imagine that Lewis and Clark had been limited to hiking in the dark of night with just a few small lights to guide them, and you'll begin to appreciate what it's like to explore the bottom of the ocean.

Meanwhile, our scientific understanding of the ocean depths has had a huge effect on geophysics, climate science, and biology. The theory of continental drift was put forth by Alfred Wegener in the early 1900s, but it was not until the 1960s, when expeditions mapped out the remnant magnetization of the seafloor, that the idea that Earth's surface is made up of moving tectonic plates gained wide acceptance. With plate tectonics came the insight that convection of heat from Earth's active mantle powered spreading centers on the ocean floor, creating ridges of seafloor mountains, volcanoes, and hot springs. Often compared to the seams on a baseball, these spreading ridges provide a new surface to the seafloor, replacing old surface material being subducted along the plate margins.

In 1977, a submersible expedition to explore the geology of a spreading ridge off the coast of Ecuador quickly turned into a biology expedition. The team had thought they might observe a few of the deep-sea creatures found in earlier dredging efforts; what they discovered was a stunning subsurface biosphere thriving off the rich chemistry emanating from hydrothermal vents positioned along the ridge. Unlike ecosystems on Earth's surface, which are powered by photosynthesis at the base of the food chain, such hydrothermal-vent ecosystems are powered by chemosynthesis: microbes eat the compounds coming from the vents and serve as the base of the food chain. This discovery sent tidal waves throughout the world of biology and has led to countless PhDs dissertations and papers focused on understanding the chemistry and dynamics of those alien biospheres. In the years since the discovery of hydrothermal vents, numerous other systems have been discovered along other spreading ridges and in areas of active undersea volcanism. Such systems fascinate those of us focused on the search for life in distant oceans, where thick, icy shells likely limit photosynthesis to marginal niches.

In retrospect, it seems unsurprising that plate tectonics would generate hot springs at the bottom of the ocean and that life would somehow find a way to thrive. At the time of the discovery, of course, the acceptance of plate tectonics was relatively new and the field of microbiology was just beginning to enter the revolution brought about by the discovery of DNA. What is surprising to this day is the discovery in 2000, by Deborah Kelley and her colleagues at the University of Washington's School of Oceanography, of an active system of hydrothermal vents where none were expected—20 kilometers west of the Mid-Atlantic Ridge. The site has come to be called Lost City, and the vents definitely invoke this name. In 2003, I had the opportunity to dive down to Lost City, a kilometer below the ocean's surface, and see the astonishing field of white carbonate towers and surrounding ecosystem with my own eyes. Flowing gently up and through the towers is a shimmering sheath of hot water, around which lurk various creatures of the deep. It's a vision I keep in my head as background for daydreams about how life began on Earth and how it might survive on worlds like Europa.

Three attributes make Lost City remarkable, both in the context of life on Earth and in our search for life on ocean worlds beyond Earth.

1. Lost City is not powered by heat coming from the

mantle. The energy of the vents at Lost City comes from an exothermic (heat-producing) reaction resulting from the interaction of ocean water with freshly exposed mantle rock—specifically, peridotite. The reaction is called serpentinization, because one of the products is serpentine, a green and somewhat scaly-looking rock. Though the details of this transformation are quite different from those of chemical hand warmers, which typically operate by an accelerated oxidation process, one can use hand warmers as an analogy. When you open a hand-warmer packet, you allow oxygen to enter, and the exothermic reaction proceeds until all the iron inside has turned to rust. Similarly, whenever cracks form in the seafloor crust, water enters, and the serpentinization reaction proceeds until all the exposed peridotite has been altered. In the process, heat, serpentine, carbonate rocks, and some biologically useful compounds are produced. All you need is fresh mantle rock and water; you don't need a hot, convecting mantle of molten rock to power this kind of system.

2. As noted, Lost City does not lie along one of the "baseball seams"—the spreading ridges—of plate tectonics. Lost City is a so-called off-axis site and represents a new geographic class of hydrothermal systems. To date, Lost City is the only known large-scale hydrothermal system of its kind, and scientists are eager to explore new regions of the seafloor in the years to come to see just how common off-axis serpentinizing systems are. The Lost City system carries important implications for hydrothermal activity on small ocean worlds of the outer Solar System, where internal heating from radioactive decay and tidal energy might not be sufficient to maintain an active mantle

and thus active plate tectonics: the usual "black smoker" hydrothermal vents found along spreading centers may not be possible. However, Steven D. Vance, my colleague at the Jet Propulsion Laboratory, has shown that cracking from cooling and tidal flexing on the seafloors of some of those small worlds could be sufficient to sustain serpentinizing systems.[1]

3. Finally, and perhaps most important, the type of geochemical environment at Lost City is one that has many researchers who study the origin of life intrigued, including me. Michael J. Russell, also at JPL, has long argued that low-temperature alkaline aqueous environments are more suitable for the chemistry that eventually leads to the origin of life.[2] Lost City provides this type of environment. Unlike the black smokers, which are acidic and can reach temperatures of 400°C, serpentinizing vents reach temperatures of only about 70°C and maintain a high, or basic, pH, between 9 and 11. This is a good environment from the standpoint of the building up of large molecules and their collection into self-contained compartmentalized units, or protocells. The chemical reactions possible at Lost City, however, are what Russell and others find most compelling. Methane is produced during serpentinization, and subsequent reactions on mineral surfaces can lead to the reduction of carbon dioxide, which then leads to the establishment of a geochemical proton gradient—i.e., ionized

---

1 S. Vance et al., "Hydrothermal Systems in Small Ocean Planets," *Astrobiology* 7 (2007), 987–1005.

2 See, for example, W. Martin, J. Baross, D. Kelly, and M. J. Russell, "Hydrothermal Vents and the Origin of Life," *Nature Reviews: Microbiology* 7 (2008), 805–14.

hydrogen (hydrogen atoms without their electron) moving across the vent-ocean interface. This proton gradient may have served as the first biochemical battery to help power life. It all amounts to the primordial metabolism that may have kick-started life here on Earth.

This final point underscores what is perhaps the most revolutionary aspect of the coming age of ocean exploration: the prospect of discovering a second, independent origin of life—life unrelated to the tree of life on Earth. With such a discovery, we immediately transform our Universe from one in which life on Earth is a biological singularity to one in which life arises wherever the conditions are suitable—a biological universe. Coupled with this is the possibility of probing whether the DNA → RNA → protein biochemical paradigm is the only game in town. To the best of our knowledge, all life on Earth utilizes the same biochemistry. Perhaps other solutions exist. Perhaps, as we explore other worlds, the range of biochemistries suitable for life will emerge as a kind of periodic table for life, with variations on the blueprint molecule (DNA), the messenger molecule (RNA), and the structural molecules (proteins) defining a biochemical landscape for life in the Universe.

Four hundred years ago, Galileo Galilei turned his telescope to the sky and revolutionized our understanding of the Universe. His discovery of the moons of Jupiter helped put the final nail in the coffin of Aristotelian cosmology— the idea that the Earth was at the center of the Universe. If Jupiter had moons, then it too must be a planet. Earth was now one of many planets orbiting the Sun, and soon the Sun was understood to be a star, one of many in the Universe. In the centuries after Galileo, our understanding of the laws of physics would be extended from the Earth to worlds and wonders beyond. So, too, with the advent of spectroscopy and planetary rovers, our understanding of chemistry and geology has extended to worlds and wonders beyond Earth.

But the science of biology has yet to make that leap. We now stand, with the coming age of ocean exploration, on new shores of discovery that could catalyze this revolution. It has been nearly 140 years since HMS *Challenger* began unveiling secrets from our ocean depths. Just decades from now, signals from our fleet of robotic explorers in distant oceans throughout the Solar System may carry news of discoveries that will once again revolutionize our understanding of the Universe in which we live.

FELIX WARNEKEN
received his PhD from the University of Leipzig while working at the Max Planck Institute for Evolutionary Anthropology. He is an assistant professor of psychology at Harvard University, where he heads the Social Cognitive Development Group in the university's Laboratory for Developmental Studies. His research focuses on the origins of cooperation in humans and chimpanzees.

CHILDREN'S HELPING HANDS

FELIX WARNEKEN

One of the key human characteristics is our tendency to act on behalf of others, by sharing such resources as money and food with people in need or comforting people in distress. As adults, we do this routinely, often without immediate personal gain and occasionally even when such behavior is costly to us. It is often assumed that such altruistic behaviors are cultural in origin: our parents taught us moral norms, say, or rewarded us for being nice to others. Moreover, many people think of these behaviors as uniquely human, holding that other animals don't act altruistically in these ways because they are driven by selfish motivations alone and don't have parents who teach them how to be an altruist.

However, several novel empirical findings suggest that

human altruism has deeper roots than previously thought. Specifically, my colleagues and I have conducted studies showing that human children act altruistically from a very early age—that is, before specific social experiences, such as being taught cultural norms, could have significantly influenced their development. Moreover, even chimpanzees on occasion act helpfully, raising the possibility that we are perhaps not as special in our altruism as we might think. By studying young children, we can determine which altruistic behaviors we're capable of early in our lives and trace the development of those predispositions as they interplay with cultural norms and moral education. And by testing chimpanzees, one of our two closest living evolutionary relatives, we can time-travel into our evolutionary past, differentiating any altruistic acts that may have characterized our common ancestor from those that emerged only in the human lineage. Comparing the behavior of young children with that of chimpanzees can thus provide answers debated since the times of the philosophers Thomas Hobbes and Jean-Jacques Rousseau: Is altruism founded in social norms adopted to keep our selfish nature in check (the Hobbesian view)? Or, as Rousseau supposed, are we naturally inclined to care about others?

Early in their lives, children are eager to find out why and how people do what they do. And children take things in with surprising sophistication. Here's an example: when one-year-olds watch someone use a novel tool or press buttons on a fancy apparatus that creates a startling effect, they can tell what the person did on purpose and what was an accident (which is often accompanied by surprise: "Whoops!"). When it's their turn to wield the tool or press the buttons, they don't copy everything the person did but only what the person intended to do. Children are intention readers, not just behavior copiers. This intention-reading capacity comes in handy: When children learn by observing others, they separate the wheat from the chaff and imitate only those aspects of another person's behavior worth copying.

What occurred to me was that another domain in which intention reading is essential is helping. In order to help someone with a problem, the helper has to be able to identify what the person is trying but failing to achieve. Would young children use their intention-reading capacity not only for their own ends (How does this tool work? Which button makes the TV turn on?) but also to help others? For instance, when someone drops something and reaches for it, will they understand that the dropping was an accident and the other person is now trying to pick the object up? Will they help? The opportunity offered itself when I was testing a one-year-old boy in a study on social play, crawling on the floor with him so as to be an appropriate play partner (I am six feet, six inches tall). At one point a ball accidentally rolled out of my reach and I pretended to be unable to retrieve it, stretching awkwardly across the floor. And indeed, the boy stood up, picked up the ball, and put it in my hand.

This serendipitous moment inspired a suite of studies investigating altruistic behavior in young children. What became apparent from these studies is that children help others in various ways and begin doing so early in life. With Michael Tomasello from the Max Planck Institute for Evolutionary Anthropology in Leipzig, I created several situations in which eighteen-month-old children observed an experimenter performing an action when suddenly a problem occurred that prevented him from achieving his goal.[1] We found that the children helped spontaneously, without being asked, receiving a reward, or being praised for their efforts. They picked up clothespins an experimenter had dropped on the ground and was unsuccessfully reaching for. They opened the doors of a cabinet when the experimenter bumped into it while carrying a stack of magazines he was trying to put inside. They helped put a book back on top of a pile after it had slipped off. After they'd learned that a certain box could be opened by lifting a flap and they saw the experimenter accidentally drop a spoon into the box through a hole and squeeze his hand through the hole in a vain attempt to retrieve it, they used their newly acquired technique to open the box and get the spoon for him.

It's important to note that our eighteen-month-old subjects did not perform these behaviors in control conditions, in which the same basic situation was established but with no indication that it presented a problem for the experimenter (e.g., he threw a clothespin on the floor on purpose, or the cabinet doors were closed but he was trying to put the magazines on top of the cabinet rather than inside). This ruled out the possibility that they were acting without regard to the other person's need—just because, say, they liked to hand things to adults or

1 F. Warneken and M. Tomasello, "Altruistic Helping in Human Infants and Young Chimpanzees," *Science* 311 (2006), 1301–3.

liked to open cabinet doors. Our subjects seemed able to determine whether help was needed or not and could do so in a variety of situations, exhibiting the sophisticated intention-reading capacities that emerge early in childhood.

Young children are also willing to put some effort into helping. Further studies showed that they continue to help over and over again, even if in order to pick up a dropped object they have to surmount an array of obstacles or stop playing with an interesting toy. We had to be inventive in creating distracting toys that might lower their tendency to help—flashy devices that lit up and played music; colorful boxes that jingled when you threw a toy cube into them and shot it out the other end. We decided that if we couldn't sell the scientific community on our results, we could at least go into the toy business.

As noted, the behavior of our little subjects did not seem to be driven by the expectation of praise or material reward. In several studies, the children's parents weren't in the room, and thus the helping cannot be explained by their desire to look good in front of Mom. In one study, children who were offered a toy for helping were no more likely to help than those children who weren't. In fact, material rewards can even have a detrimental effect on helping: During the initial phase of another experiment, half the children received a reward for helping and the other half did not. Subsequently, when the children again had the opportunity to help but now without a reward being offered to those in either group, the children who had been rewarded initially were less likely to help spontaneously than the children from the no-reward group.[2] This perhaps surprising result suggests that children's helping is intrinsically motivated rather than driven by the expectation of material reward. Apparently, if such rewards are offered, they can change children's original motivation, causing them to help only because they expect to receive something for it.

These studies demonstrate that very soon after their first birthday, young children begin to behave altruistically, performing acts of what we have named "instrumental helping," in which they infer another person's unfulfilled intention and help to bring it about. These results square with studies showing that, at around the same age, children begin to act on others' behalf in various other ways. For example, one- and two-year-olds will comfort someone in distress, showing a capacity for resonating with the emotional states of others. Moreover, when children begin to point, at about one year of age, they use this newly acquired nonverbal device not only to request objects from others (the so-called imperative point, aka "Give me that!") but also to help someone who is looking for something ("There it is!").[3] Two-year-olds will on occasion share objects with others—although often only after the person explicitly states her wishes. Not surprisingly,

---

2 F. Warneken and M. Tomasello, "Extrinsic Rewards Undermine Altruistic Tendencies in 20-Month-Olds," *Developmental Psychology* 44 (2008), 1785–8.

3 U. Liszkowski et al., "Twelve- and Eighteen-Month-Olds Point to Provide Information for Others," *Journal of Cognition and Development* 7 (2006), 173–87.

they're more reluctant to share their own belongings than other, less valuable objects.[4]

Taken together, these studies in the field of developmental psychology demonstrate that young children are not oblivious to the needs of others. In addition to all the self-focused and selfish things children do, they can act on behalf of others if the occasion arises. The fact that these behaviors emerge so early in children's lives is important, because it suggests that the social and moral norms of one's culture are not the original source of the emergence of altruistic behaviors in humans. It can still be argued that young children are especially quick social learners—or that parents who are particularly motivated to raise altruistic offspring will use subtle socialization techniques that developmental psychologists have not yet detected. However, there is another empirical approach that can be instructive: comparing the behavior of human children to that of chimpanzees.

Studying chimpanzees to learn about child development might seem a bit far-fetched. So let's step back for a second and ask, What can we learn from chimpanzees about human behavior? One thing we can learn is whether certain cognitive capacities and social practices are a necessary prerequisite for other types of behaviors. For example, chimpanzees can discriminate between and remember small quantities, such as one versus two or three versus four. When you show them a pile of three grapes and a pile of four grapes, and then conceal the three grapes in

---

4 M. Svetlova et al. "Toddlers' Prosocial Behavior: From Instrumental to Empathic to Altruistic Helping." *Child Development* 81(6) (2010), 1814–27.

one cup and the four grapes in another, they will consistently choose the cup with four grapes. Chimpanzees do this even with no linguistic abilities and without having been trained to use symbols; doing basic math thus doesn't require knowing the words "one, two, three" or the mastery of the mathematical symbols 1, 2, 3 (let alone going to math class). Something else we can learn from chimpanzees is what our evolutionary ancestors were like. Humans, chimpanzees, and bonobos evolved from a common ancestor, reflected in the fact that we share most of our genetic material—about 98 percent of it. Between 5 million and 7 million years ago, some individuals branched off and went their own way, their descendants eventually becoming modern humans (chimpanzees and bonobos became separate species only about 1 million years ago). The assumption is that behaviors shared across these species also characterized our common ancestor, whereas behaviors that only humans express evolved after the great split and are thus human-specific.

So are our altruistic tendencies unique to humans, or do chimpanzees share some of them? If cultural practices—such as internalizing social norms or being taught how to behave—are the main source of our altruistic behaviors, we would not expect to see those behaviors in chimpanzees. This is because, to the best of our knowledge, chimpanzees don't teach their children how to behave toward other individuals, nor do they enforce communally shared cultural norms. The absence of these socialization practices suggests the hypothesis that only humans develop a motivation to act on behalf of others, whereas chimpanzee behavior is guided solely by selfish interests leading to personal gain.

Several experiments with chimpanzees show exactly that. One of the telling experiments performed by a research team headed by a UCLA anthropologist, Joan Silk, works like this: A chimpanzee subject is given the opportunity to pull a handle that moves a container of food toward it and simultaneously moves an empty container toward another chimpanzee (the 1/0 option). Alternatively, the chimp can pull a handle that moves food to both of them (the 1/1 option). In Silk's study, the subject chimpanzees did not preferentially pull the handle that also delivered food to the recipient chimpanzee, even though that gesture could be made at no cost to them. They appeared indifferent to the outcome for the other, at least in this context, leading to the conclusion that chimpanzees are indifferent in general to the needs of their conspecifics.[5]

However, when we tested chimpanzees in instrumental helping situations similar to the tests with human toddlers described above, we were amazed to see that chimpanzees, too, proved helpful at times. At the Leipzig Zoo, we conducted a study with human-reared chimpanzees, who observed their human caregiver accidentally drop an object and unsuccessfully reach for it (just like the clothespin-dropping test with the human children). The chimpanzees picked the dropped objects up and brought them to the caregiver when she was reaching for them, but not in control conditions, in which she showed no inter-

---

5 J. Silk et al., "Chimpanzees Are Indifferent to the Welfare of Unrelated Group Members," *Nature* 437 (2005), 1357–59; K. Jensen et al., "What's in It for Me? Self-regard Precludes Altruism and Spite in Chimpanzees," *Proceedings of the Royal Society of London B* 273 (2006), 1013–21.

est in the dropped objects. They performed this behavior even though she did not reward them for their help. This was the first experimental demonstration of helping in chimpanzees.[6]

An important concern, of course, was that this kind of helping might be restricted to human-reared chimpanzees interacting with their primary (human) caregiver: Maybe they wanted to please her, since she had already trained them to do all sorts of things at her command. So my colleagues Brian Hare and Alicia Melis went to a chimpanzee sanctuary in Uganda to test individuals with a different rearing history, who, moreover, had never interacted with those two researchers. We found that those chimpanzees, too, helped by fetching an object the experimenter was reaching for—and significantly more often than in situations in which the experimenter was *not* reaching for it.[7] Even more surprising was that the offer of a reward had no effect on this behavior. Chimpanzees picked up the objects for the experimenter because the experimenter wanted them, not because they wanted a reward for themselves. So far, so good, but the critical test case would obviously be one that showed whether chimpanzees help other chimpanzees.

To test this, we confronted chimpanzees with the following situation: A recipient chimpanzee tries to open a sliding door to get into a room where food is waiting for

---

6 Warneken and Tomasello, "Extrinsic Rewards."

7 F. Warneken, B. Hare, A. P. Melis, D. Hanus, and M. Tomasello, "Spontaneous Altruism by Chimpanzees and Young Children," *PLoS Biology* 5 (2007), 1414–20.

her. However, the door is blocked by a chain attached with a peg to the bars in an adjacent room. We found that when we placed a chimpanzee subject in the adjacent room, frequently it would release the chain so that the recipient chimpanzee could get the food. This behavior occurred less often in control conditions, in which either the recipient was trying to go through another door or no recipient was present. Thus, these chimpanzees seemed able to determine when another needed help and respond accordingly.

A major challenge for the future is to find out under what other circumstances chimpanzees show altruistic tendencies—and under what circumstances they do not. In several studies, chimpanzees were somewhat reluctant to provide food actively to others, even at no cost to themselves. On the other hand, we had demonstrated that they do intervene on another's behalf in instrumental helping situations. What explains this discrepancy? One pattern that appears to emerge is that chimpanzees help only when the helpee overtly expresses the need for help: reaching for a desired object, trying to open a door, or gesturing toward the potential helper. When these signals are absent, chimpanzees do not proactively engage in altruistic behaviors. This might reflect limitations in reading another's intentions or simply indicate a generally weaker altruistic motivation—that is, the helpee needs to work harder to persuade its conspecifics to provide assistance. Our tests showing altruistic chimpanzee behavior (in the sense of intervening to further another's goal) did not involve any sacrifice of resources or major effort on the part of the helper; thus, it remains an open question whether chimpanzees will engage in altruistic behaviors that come at a cost. It is possible that they are willing to act altruistically if doing so is fairly cheap but are less prone to act altruistically when it isn't. What we can say at this point is that altruistic tendencies are not absent in chimpanzees. Also, they seem to have the fundamental cognitive and motivational capacities for engaging in altruistic behavior. This suggests that human altruism might have evolutionary roots dating back at least to the last common ancestor of humans and chimpanzees.

These findings also help us better understand the factors responsible for the emergence of altruistic behaviors in human children. Altruistic behaviors do not appear to be solely the outcome of cultural norms and socialization. No doubt socialization practices profoundly influence children's development and cultural norms can facilitate and sustain whatever is jump-started by biological inheritance. Despite altruism's early emergence, children have a lot to learn about how and whom to help. For example, we should not blindly help everyone (such as those with bad intentions); it is essential that we learn when to help and when not to help. Moreover, our altruistic tendencies become an important asset; being helpful tends to improve your reputation, whereas failure to help others can damage it. You need sophisticated perspective-taking abilities to know how your behavior affects how others see you—a strategic skill that young children lack but that becomes important later in life. It remains to be studied at what stage of their development children begin to take into account the complex intricacies and norms that characterize our social life—and human altruism in particular. However, adherence to cultural norms does not appear to be the original source of our altruistic behaviors. Rather, it appears that cultural factors can build on a biological predisposition we share with our closest evolutionary relatives: Culture cultivates, rather than implants, the propensity for altruism in the human psyche.

WILLIAM McEWAN
is a virologist working on intracellular immunity to viruses. His research focuses on specific mechanisms in mammalian cells that actively overcome viral infection. He graduated with a BSc in genetics from University College London in 2005 and did a master's degree and PhD at the University of Glasgow, researching immunity to lentiviruses in cats and lions. He is currently a postdoctoral researcher at the MRC Laboratory of Molecular Biology, Cambridge, U.K., where he continues to probe the biology of antiviral immunity.

MOLECULAR CUT AND PASTE

THE NEW GENERATION OF BIOLOGICAL TOOLS

WILLIAM McEWAN

This afternoon I received in the post a slim FedEx envelope containing four small vials of DNA. The DNA had been synthesized according to my instructions in under three weeks, at a cost of 39 U.S. cents per base pair (the rungs adenine-thymine or guanine-cytosine in the DNA ladder). The 10 micrograms I ordered are dried, flaky, and barely visible to the naked eye, yet once I have restored them in water and made an RNA copy of this template, they will encode a virus I have designed.

My virus will be self-replicating, but only in certain tissue-culture cells; it will cause any cell it infects to glow bright green and will serve as a research tool to help me answer questions concerning antiviral immunity. I have designed my virus out of parts—some standard and often used, some particular to this virus—using sequences that hail from bacteria, bacteriophages, jellyfish, and the common cold virus. By simply putting these parts together, I have infinitely increased their usefulness. What is extraordinary is that if I had done this experiment a mere eight years ago, it would have been a world first and unthinkable on a standard research grant. A combination of cheap DNA synthesis, freely accessible databases, and our ever-expanding knowledge of protein science is conspiring to permit a revolution in creating powerful molecular tools.

Nature is already an expert in splicing together her existing repertoire to generate proteins with new functions. Her unit of operation is the *protein domain,* an evolutionarily independent protein structure that specializes in a particular task, such as an enzymatic activity or recognition of other proteins. We can trace the evolutionary descent of the protein domains by examining their sequences and grouping them into family trees. We find that over the eons of evolutionary time the DNA that encodes protein domains has been duplicated and combined in countless ways through rare genetic events, and that such shuffling is one of the main drivers of protein evolution. The result is an array of single- and multidomain proteins that make up an organism's proteome. We can now view the protein domain as a functional module, which can be cut and pasted into new multidomain contexts while remaining able to perform the same task. This modular capability immediately lends itself to engineering: We don't have to go about finding or artificially evolving a protein that performs our chosen task; we merely combine components that together are greater than the sum of their parts.

I'm interested in the defense mechanisms within cells—mechanisms that specifically recognize and disable intracellular pathogens. This type of defense is considered separate from the two main branches of immunity that are more intensely studied: the evolutionarily ancient "innate" immune system and the vertebrate-specific "adaptive" immune system. Innate immunity is the recognition of conserved features of pathogens—for example, the detection by specialized cells, such as macrophages, of the sugary capsule that surrounds many bacteria. Adaptive immunity works by fielding a huge diversity of immune recognition molecules, such as antibodies, and then producing large quantities of those that recognize nonself, pathogen-derived targets. The newly discovered kind of immunity on which I work, sometimes termed "intrinsic immunity," shares features with innate immunity but tends to be widely expressed, instead of residing just within "professional" immune cells, and is always "on." In other words, every cell in an organism is primed and ready to disable an invading pathogen. The intrinsic immune system is at a strategic disadvantage, as its targets are often fast-evolving viruses that can rapidly mutate to evade recognition. Unlike the adaptive immune system, which can quickly generate a response to an almost infinite diversity of targets, the intrinsic immune system must rely on rare mutations and blind selection over evolutionary time to compete with its opponents.

So far, the study of the intrinsic immune system has been dominated by its interaction with retroviruses. The retroviruses, an ancient affliction of vertebrates, violate the central dogma of biology—that DNA makes RNA makes protein[1]: they are RNA viruses able to generate DNA copies of themselves and insert this Trojan-horse code into the host's genome. Almost one-tenth of the human genome is the defunct relic of this sort of infection.

Within the past 7 million to 12 million years, a comparatively recent member of the retrovirus family, the lentivirus, has emerged and spread slowly through the branches of the mammalian family tree. The oldest known traces of lentivirus have been found in the genome of rabbits, but current infections occur in horses, cats, ruminants, and primates. Lentiviruses arrived in humans in the form of HIV, as several cross-species transmission events from other primates. Only one of those viral transmissions—from chimpanzee to human, sometime in the late nineteenth or early twentieth century—has adapted to its new host in such a devastating manner, the virus being HIV-1 M-group, which causes AIDS and currently infects 33 million people worldwide.

One of the major players in intrinsic immunity is TRIM5, a four-domain protein that is expressed in virtually every cell in the human body. By virtue of one of its domains—the RING (which stands for Really Interesting New Gene) domain—TRIM5 has an enormously high turnover rate; that is, each of its molecules is degraded within about an hour of the cell's having synthesized it. By virtue of another of its domains, it can recognize and

---

1 This commonly used shorthand for the central dogma is not actually what Francis Crick intended; he wanted to imply that information was unable to flow back to nucleic acid once translated into protein. This more precise version of the central dogma is not violated by retroviruses.

engage retroviruses soon after their entry into the cell. As a result, incoming viruses can be degraded along with TRIM5 and thus made noninfective. A classic arms-race situation has developed, wherein TRIM5 has tried to maintain its ability to recognize the rapidly evolving retroviruses, placing the gene under some of the strongest Darwinian selection in the entire primate genome. However, HIV-1 seems to have the upper hand at the moment: the human TRIM5 variant only marginally reduces the replication of HIV-1. Could this be one of the failures in human immunity that has permitted such a dramatic invasion by this pathogen? And what does human TRIM5 need to do in order to gain the upper hand? Or, to ask a bolder question, what can we do to it to engineer resistance to the disease?

One surprising answer is provided by protein-domain fusions in other primate species. A fascinating thing has happened in South American owl monkeys: TRIM5 has been fused with a small protein that HIV-1 depends on for optimal replication. The resulting fusion protein is called TRIMCyp and can reduce the replication of lentiviruses by orders of magnitude, essentially rendering the owl monkeys' cells immune to the virus. Almost unbelievably (and it amazes me that espousers of Intelligent Design aren't onto this), this feat of genomic plasticity has happened twice: versions of TRIMCyp have also been described in the unrelated macaque lineage. Since no wild populations of owl monkeys or macaques have been shown to harbor lentiviruses, it is difficult to say whether TRIMCyps have been selected specifically to combat lentiviruses, but there remains the intriguing possibility that TRIMCyp has helped lead to the current lentivirus-free status of one or both of these species.

So how can we benefit from gene fusions in other species? The first lesson is that by splicing together domains from seemingly unrelated proteins, unexpected and useful products can be generated. Researchers have already generated human TRIMCyps, which would avoid the immune-system rejection that introducing an owl monkey gene would produce. My colleagues and I have also engineered a feline TRIMCyp that prevents replication of the feline immunodeficiency virus in tissue-culture systems. However, in a clinical setting TRIMCyp must be expressed within cells to be useful as an antiviral, and the only effective means of achieving this is through gene therapy to alter the target cell's genetic material. In a neat twist of roles, the best means we have of doing this is with a modified retroviral vehicle, or vector, to introduce a stretch of engineered DNA into the genome.

Two years ago, the first person in history was rid of an HIV infection. The situation was unusual, as the patient also suffered from leukemia, meaning that a bone marrow transplant was required. Instead of choosing a typical donor, an individual with a rare but naturally occurring mutation that prevents HIV entry into cells was chosen. The donor's mutation impairs the protein CCR5, a cell-surface receptor that HIV requires for entry. The mutation has helped inspire some of the most recent antiretrovirals, CCR5 inhibitors. The recipient of this "delta-32" mutant CCR5 marrow showed no sign of HIV replication eighteen months after the transplant, despite having ceased antiretroviral treatment. Using this example as a proof of principle, it is quite possible that within the next five years a similar treatment will be delivered to a leukemia/HIV patient but with human TRIMCyp providing the antiviral activity. Indeed, we already have the two naturally occurring animal models—owl monkeys and macaques—that suggest that the fusion is relatively benign. Farther down the line, it might be possible to alter the genetic makeup of the target cells without the need for a bone-marrow transplant, a prophylactic gene therapy that could be as easy to deliver as any other vaccine. Given the issues surrounding the public acceptability of genetically modified crops and even conventional inoculations in some circumstances, whether such treatment will prove to be ethically and socially acceptable remains to be seen—but the technology is fast approaching, and we will have to decide whether or not to take advantage of it.

Of course, the power of fusing domains is not limited to antivirals. The scope of this technique is almost limitless, and the first generation of protein-fusion therapeutics is already upon us. Many of the recent anticancer drugs are protein-domain fusions—mostly fusions between mouse and human antibodies that block aberrant pathways specific to the cancer in question. The fusion is required because the mouse antibodies must be "humanized" in order to prevent an immune reaction to the foreign protein. Another potentially exciting and recently widely publicized development is a gene-therapy vector based on the adenovirus, which causes upper respiratory infections. Researchers can simply drop in "cassettes" of DNA to create fusion proteins that retarget the virus to specific cancer cells, where it will replicate and kill its target. There will also be nonmedical biotechnology applications for fusion proteins—for instance, the modification of enzymes to perform particular tasks, such as hydrogen or biofuel generation. Applications may be even more far-fetched: I recently attended a lunchtime seminar where arrays of viruses with domain fusions were being touted as electrical components of the future.

Using traditional techniques (and in molecular biology, "traditional" refers to techniques of the past twenty years), you can generate such fusions once you have your hands on a physical copy of all the genes you want to splice together and a student or postdoc, such as myself, to carry out the necessary protocols. With luck, it's a week's work to make a fusion or two; without luck (or competence), it can take a month. Getting access to the desired DNA often involves a legal material-transfer agreement, which can add further weeks to the procedure. And because time is money, we are fast approaching the moment when generating the fusion by hand is more costly than having the gene synthesized at facilities that specialize in such work. The cost of synthesis is dropping fast. Since I placed the order for my virus, the best quote for gene synthesis for a small gene has dropped from 39 to 35 cents per base. However, the economic case alone massively undersells what gene synthesis has to offer molecular biology. We need only the *sequence* information, not the physical DNA template itself, and a researcher can select any sequence in the public domain. Suddenly, with only an Internet connection, a hydrogenase from a sea-dwelling archaebacterium is just as obtainable as lion TRIM5. The new central dogma then becomes Silicon Makes DNA Makes Something Useful.

There will never be an exclusive field of research called domain-fusion science. The fusions will be made by researchers, clinicians, and engineers who in their own line of work require a protein with the function of more than one existing domain. Some of the fusions will be obvious next steps, some will be inspired leaps, and some will even be pulled from randomly assembled large-scale screens of domain fusions. The closest field in its own right to "domain-fusion science" is synthetic biology, a discipline that is busy systematically cataloging and defining the parts necessary to engineer useful biological systems. This includes domain fusions but does not stop there. The parts include the regulatory "control switches" of genes— promoters, enhancers, and repressors—as well as suites of genes that are designed to act in a tractable and predictable manner. In essence, synthetic biology is the reduction of complexity to modularity in order to rebuild complexity. To many researchers, it will seem a statement of the obvious even to note the existence of domain fusions, yet fusions have provided some of the most innovative therapeutics and molecular tools of recent decades, and their effects will be increasingly felt in the years to come. Their application in immunity really excites me. There has always been a twin requirement to recognize and then disable pathogens. If we can create therapeutics that at once engage and label the invaders as threats in a manner comprehensible to the cell, we may well regain the advantage over retroviruses and defeat many of our other current afflictions. As for my synthetic virus, it now remains for me to see if it works.

ANTHONY AGUIRRE
holds a BS (1995) in mathematics and physics from Brown University and a PhD (2000) in astronomy from Harvard University. He is an associate professor of physics at the University of California, Santa Cruz, where he studies a variety of topics in theoretical cosmology, including the early universe and inflation, gravity physics, first stars, the intergalactic medium, galaxy formation, and black holes.

NEXT STEP

INFINITY

ANTHONY AGUIRRE

The question of whether the world is finite or infinite has bedeviled us for a long time. It was a classic question in ancient Indian philosophy. Aristotle cogently argued that while infinity made sense in the "potential," the world could not "actually" be infinite. Giordano Bruno declared the world infinite and was burned at the stake. Galileo, more circumspect, had his clever alter ego, Salviati, completely befuddle Simplicio with how paradoxical and slippery infinity is. And Immanuel Kant really threw down the gauntlet, arguing that both an infinite and a finite world were logically impossible: an infinite universe would take an infinite time to be "synthesized" and thus could never at any one time be said to be infinite—but a finite universe must somehow be embedded in a seemingly meaningless "emptiness" that is not part of the universe.

Because finite and infinite spaces alike tax our conception of space, and because we, as finite creatures, clearly cannot measure or directly observe an infinite system, it might appear that the question could most conveniently be consigned to the dustheap of purely philosophical inquiries that hard-nosed scientists can safely ignore.

Yet Albert Einstein's theories of space and time radically reformulated the questions of finite and infinite spaces and times, and the ensuing development of cosmology has brought infinity into the domain of testable physical science. For example, a uniform space can be curved like a sphere—and comprise a universe that is finite in volume without having any "edge" or empty space outside it. Even more impressive are the tricks that relativity can play concerning infinite spaces, which have come to occupy a central place in contemporary cosmology. To tell this story, I will contend in the following four sections of this essay that:

1. It is logically possible for a finite spatial region to evolve into a space that is both uniform and spatially infinite.
2. Inflation, our well-tested and widely accepted paradigm for understanding the very early universe, provides just the right physics and context for this to happen.
3. In the combination of our current best bets for cosmology and fundamental physics, inflation creates infinitely many "subuniverses," each itself uniform and infinite.

4. This raises profound issues—for physics, cosmology, and even personal identity—that we cannot ignore.

1. *A spatial region can evolve into an infinite, uniform space.*

While this idea appears absurd, the apparent absurdity is based on intuitions about space and time that were profoundly undermined in 1905 by Einstein's special theory of relativity.

A fundamental lesson of relativity theory is that there is no single objective definition of what events are happening "now" across a large region. An observer can easily categorize nearby events into those that have happened, those occurring "now," or those that might happen in the future. Yet faraway events cannot be immediately seen; information about them travels with, at most, the speed of light, so when a faraway object is observed, it is seen as it was some time ago. An observer must therefore infer, using a careful procedure, which events happened at the same time as a given event happening to the observer. What Einstein showed is that if two observers do this, and if the observers are moving with respect to each other, then they will make *different inferences*. According to one observer, two faraway events might happen at the same time; yet according to another observer, whose view is just as valid as the first, they happen at different times. More generally, given two events, different observers will ascribe different spatial distances, and different temporal intervals, between the two events (see Figure 1). What all observers will agree on, according to relativity, is the speed of light. This means

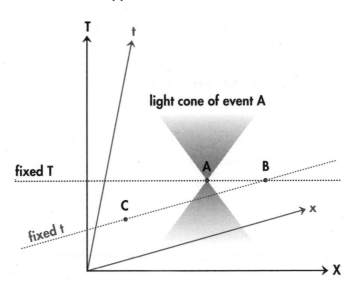

FIGURE 1: *A space-time diagram of one spatial and one temporal dimension, in which light travels on diagonal paths. A given observer can label events by spatial coordinates X and times T, and describe all events at a given time T (shown as the "fixed T" line and including events A and B) as happening simultaneously. Yet a second observer, in motion with respect to the first, will label the same events by a different spatial coordinate x and time coordinate t, with all events at a given time t (indicated by the "fixed t" line) described as occurring simultaneously. Two events at the same time t (such as A and C) will not necessarily be at the same time T; that is, events such as A and B described as simultaneous for one observer will be described as happening at different times according to another observer. Both descriptions are equally valid. What all observers will agree on, however, is the speed of light. Thus if, for a given event A, we consider all regions of space-time that might communicate with A at sub–light speed, it constitutes the interior of a "light cone" (shaded region indicated) that all observers agree on. An event in the light cone of A can be in A's past or future, but not simultaneous with A.*

that all observers will agree on whether or not an event is inside the light cone of another event (meaning that one event can influence the other event without transmitting information faster than light speed).

In relativity, then, space and time are interconvertible and should really be combined into "space-time." To ask what "space" is like is really to ask what space-time looks like at a particular time. Yet special relativity says that space-time can be decomposed in many equally valid ways. According to Einstein's general theory of relativity (1915), in fact, any two occurrences can be said to happen at the same time as long as a signal (or other causal influence) cannot be sent from one to the other—i.e., as long as one event is not within the other event's light cone. Since special relativity forbids such influences from traveling faster than light, this criterion is equivalent to saying that the two events are outside of each other's light cones.

As a fascinating and pertinent example of this freedom, consider plain, empty space-time—of just the sort we intuitively imagine: uniform, infinite, and uncurved, with the space looking the same in every direction and at all times. Yet we can divvy up this space-time in another very interesting way. We are free to define all events on a hyperboloid (see Figure 2) as occurring at the same time (even though this includes events that would, in the usual view, be considered as occurring at different times). This is allowed because no event on this surface can send a signal to any other event on this surface, as indicated by the light cones drawn in Figure 2. Moreover, each such hyperboloid can go on forever, without ever violating this constraint; thus, each comprises an infinite space. What's less

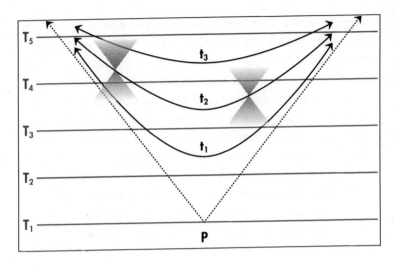

FIGURE 2: *A slice through one spatial dimension (plus time) of conventional space-time, with times indicated by gray lines and labeled by different values of T. At each time T_1, T_2, etc., space is infinite, uniform, and uncurved. Yet this space-time can with equal validity be decomposed into spaces—i.e., sets of events that happen at the same time—that consist of hyperboloids, as indicated by the black hyperbolas in the figure. Note that the light cone of any point on one of these hyperbolas does not intersect any other point on the same hyperbola. At each time t_1, t_2, etc., space is infinite and uniform but curved. In addition, all of these curved spaces fit inside the future light cone of a single point P.*

clear, but true if you go into the mathematics, is that each such space is uniform: Just as flat space has constant zero curvature, or a sphere has constant positive curvature, hyperbolic space has constant negative curvature. Finally, what is really interesting about this construction is that the whole thing is contained within the *future* light cone of a single point. This point can therefore send a signal to anywhere in this infinite space without that signal's ever exceeding light speed.

This construction directly undermines Kant's argument; even though the space at any time t is infinite, one could imagine transmitting "assembly instructions" to the whole space specifying what properties it should have. And as we'll see, this speculation is not an idle one.

2. *Our best cosmological theories provide just the right physics and context for an infinite space to come into being through a physical process.*

In order to apply his general theory of relativity to the universe, Einstein made the enormous assumption—beautifully justified since—that the universe on very large scales is essentially uniform. This allows only three geometries: the uniformly curved flat, spherical, or hyperbolic spaces mentioned above. Combining this with Edwin Hubble's observation of cosmic expansion led to the Friedmann model of the universe: an expanding space of constant curvature. This model is the foundation of Big Bang cosmology, which correctly predicts the detailed evolution of the cosmos (reckoned to be 13.7 billion years old) from a simple initial state as a hot, nearly featureless medium of (mostly) hydrogen, helium, and dark matter.

Yet that initial state is itself puzzling in several ways. Perhaps most troubling is its uniformity. Uniformity in nature generally results from some smoothing process in which variations are evened out. Yet when we observe the early universe, we can see that it is uniform over scales far larger than such processes—which are limited by the speed of light—could possibly have operated across, if the Big Bang model held all the way back to a putative "beginning of time."

Thus, a fix was devised. In the 1980s, cosmologists developed a theory called inflation, in which the very early universe expanded rapidly and exponentially, doubling in size scores of times in a tiny fraction of a second. This sort of expansion can be driven by so-called vacuum energy contained in empty space and has several crucial features. First, because space is expanding so quickly, any matter present is diluted to very low density and any waves are stretched out to form uniform fields; thus, the content of the universe looks homogeneous. Second, a small uniform region can expand so fast that the rate of change of the size of the region exceeds the speed of light. (This does not violate special relativity, which forbids signals traveling faster than light with respect to space-time itself but does not forbid objects from *separating* faster than light.) Third, positions far enough apart in inflating space cannot contact each other; because the intervening space is expanding so quickly, even a light signal sent directly from one location toward another will never arrive, as long as inflation continues. This defines a "horizon" about a given position, from beyond which no influence can be felt.

These three aspects are what allows inflation to ameliorate the peculiarities of the Big Bang's early state: An initially inhomogeneous region can be smoothed out and made clean by inflation, then expanded into a uniform region larger than our observable universe—and nothing beyond this uniform region's horizon can mess things up.

They also happen to be exactly the effects necessary to create an infinite space from a finite one.

A version of inflation in which this idea is particularly clear is called "open inflation." In this model, the universe is permeated by a scalar field known as an "inflaton"—a field that determines the density of vacuum energy at each space-time point. With high field values, the vacuum energy is large and drives rapid inflation; with low field values, the energy is low and inflation absent. The normal tendency of such a system would be to evolve from high to low energy, so that inflation would proceed for a while, then end. Imagine, however, that there is an energy barrier to this evolution. The field then gets "stuck" at large values and, as it turns out, only occasionally creates a small bubble in which the field is small. This is physics analogous to that of a carbonated beverage, which is "stuck" in the cola phase but occasionally creates small bubbles in the "bubbly" phase. The key differences are that in open inflation the bubbles expand at essentially the speed of light and the space between the bubbles also expands, even faster than the bubbles themselves.

This situation is depicted in Figure 3. An initially finite region begins to inflate. Soon thereafter, a bubble forms within it and starts to expand at a rate rapidly approaching the speed of light. Yet this bubble expansion cannot catch up with the receding edge of the initial region. Inside the wall of the bubble, the inflaton field relaxes sequentially to lower and lower constant values. Indeed, each of the surfaces in Figure 3 is just the same sort of hyperboloid as the strange dicing of conventional space shown in Figure 2. Just as in that construction, each surface constitutes

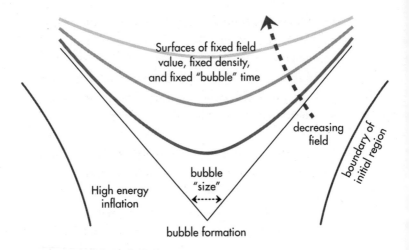

FIGURE 3: *The structure of an inflationary "bubble." Consider an initial inflating region that subsequently expands, as indicated by the outermost gray lines labeled "boundary of initial region." Within this region a bubble can form, with the border of the bubble growing at the speed of light and thus tracing out a light cone in space-time, as shown by the diagonal lines. Inside this light cone, surfaces upon which the field takes constant values are shown as hyperbolas. Like those in Figure 2, these hyperbolas constitute possible surfaces of constant "bubble" time—but here they are also surfaces of constant density, so in this description the space is of uniform density. As as this "bubble" time increases, the field evolves from high to low energy and the density decreases, until the (possible) end of inflation and the creation of matter and energy that may be termed the Big Bang.*

an infinite, homogeneous space. As such, it is natural to identify these surfaces with fixed times at which the space is progressively less and less curved—as if expanding. In short, in one way of dividing space-time into space and time, the bubble is of finite size and inhomogeneous inside; in another way, the interior of the bubble is infinite, uniform, and expanding.

The scenario depicted in Figure 3 is worth studying, because it is probably contemporary cosmology's best guess as to how our observable universe formed.

3. The creation of an infinite space probably actually happens—in fact, an infinite number of times.

Consider a small, uniform region suffused with high values of the inflaton field. It will begin to inflate and eventually spawn one bubble, then another, and another. Because the bubbles cannot grow faster than the region from which they were spawned, it can be shown that inflating space endures forever. Despite the bubbles taking "bites" out of it, the space's physical volume increases exponentially for all time. This process has been dubbed "everlasting" or "eternal" inflation, and it provides a rather different cosmic picture. The Big Bang is not the "beginning of the universe," just the end of our particular universe's inflation. And whereas the Big Bang appears to occur everywhere at once from the bubble's inside perspective, from the outside, the Big Bang—and all that evolves from it—looks like just one of many bubbles that form a sort of "steady state" in the vast, eternally enduring, inflating background.

Is this just a speculative possibility that is allowed by the physics we understand, or do we have reasons to think it actually happened? In fact we do, and they are pretty good reasons.

Although "bubbly" eternal inflation has been described

here, there are many different inflation scenarios with different properties and varying levels of connections to other fundamental theories. Most of them are eternal. Beyond explaining puzzling features of the standard Big Bang model (such as the uniformity of space), inflation makes certain predictions, which have been verified. As a recent and impressive example, seven years of data from the Wilkinson Microwave Anisotropy Probe—which looks at the Big Bang's omnipresent remnant radiation, known as the cosmic microwave background—has confirmed in great detail inflation's predictions for the particular type of nonuniformities in the early universe. Moreover, string theory, our best current candidate for a true fundamental theory, seems to require eternal inflation. Most string theorists now believe that while the four fundamental forces of nature (electromagnetism, the strong and weak nuclear forces, and gravity) are unified at the highest energies, at low energies they can be separated, but in many different ways. Just like the inflating and noninflating phases of the bubble model, these phases can coexist. Some would drive inflation and others would not, and there can be transitions between the different phases, often manifesting as, yes, bubble formation.

In this picture, dubbed the string theory "landscape," inflation brings into being an infinite number of subuniverses with a diverse set of properties spanning the array of ways in which low-energy physics can emerge from high-energy unification. Inflation can spawn bubbles inside of which is inflation, which spawns bubbles, and so on. And the incredibly rich structure of this "multiverse" is that of infinitely many times, each forming infinitely many subuniverses, each of which is infinite.

4. *This raises profound issues that we cannot ignore.*

Infinity can violate our human intuition, which is based on finite systems, and create perplexing philosophical problems. A classic example was invented by the mathematician David Hilbert. Rather than imagining an infinite universe, imagine an infinite hotel, with all rooms completely filled. Though the hotel is full, you can accommodate infinitely many more guests by moving each guest into the room of twice its current number and adding guests in all of the odd rooms. Yet although you double the number of guests, the hotel looks exactly the same. Applying this notion to an eternally inflating universe, suppose the whole bubbling mess is infinite at some time. Although infinitely many bubbles have formed during some time interval, it is rather unclear that after this time the universe is actually any bigger!

The same example reveals problems for such naive questions as "What is the chance that a randomly chosen guest is in an even-numbered room?" The seemingly obvious answer is 50 percent. Yet the number of even-numbered rooms is just as large as the total number of rooms in the hotel (since each guest found an even-numbered room to go to) and thus twice as large as the odd-numbered rooms (which make up just half of all rooms, remember?). This implies that a randomly chosen guest is twice as likely to be in an even-numbered room. Yet, by exactly the same reasoning, there are as many odd-numbered rooms as total rooms and twice as many as even-numbered rooms! Our intuition that the answer is "obviously" 50 percent arises from considering the first N rooms, then letting N approach infinity. But that's only one way of comparing the number of even- and odd-numbered rooms, and different choices—that is, different ways of measuring—would give different results.

This "counting" or "measure" ambiguity afflicts infinite systems terribly and creates a nightmare in cosmology. If postinflationary regions can have different properties, and each possible set of properties is realized in an infinite number of such regions, there is a twofold problem. First, there is no unique prediction, from this fundamental theory, for what we can observe. This is a letdown but not fatal, since we would still hope to make probabilistic predictions. Yet the measure ambiguity suggests that the relative probabilities themselves depend on the particular measure we choose and there is no compelling reason to believe any one given measure. This measure problem has spawned a large amount of recent work in inflationary cosmology, and although there has been progress, it's not clear that the progress is toward any particular resolution.

Another key aspect of infinity is that it is so much larger than anything finite. In particular, an infinite system including some randomness will tend to contain infinitely many realizations of any given finite subsystem compatible with the properties of the infinite one. This means that if we reside in an infinite bubble, then somewhere (incredibly far away) within it is another copy of the entire Earth, perfect in every detail. On a more personal level, there are infinitely many identical copies of you, as well as infinitely many of every possible small or large variation of you, some more common than others. What does this mean? If these other people are identical to you, *are* they you? What is your relation to them? If you were to suddenly cease to exist, should you take heart that "you" continue merrily along out there? Or would "you" simply find yourself to be one of them? (Beware: this possibility becomes more and more disturbing the more you think about it.)

Should we embrace the idea that our world is truly infinite, or should we look for some way to tame and regulate this infinity in our theories? It is difficult to say. What seems clear, however, is that infinity can no longer be safely ignored: beautifully constructed, empirically supported, self-consistent theories have brought infinity from idle curiosity to central player in contemporary cosmology. And if correct, the worldview these theories represent constitutes a perspective shift unlike any other: in comparison to the universe, we would be not just small but strictly zero. Yet here we are, contemplating—if not quite understanding—it all.

DANIELA KAUFER
is an assistant professor in the Department of Integrative Biology at the University of California, Berkeley. She also holds appointments in the Stem Cell Center and the Neuroscience Institute. Her research focuses on the molecular events that underlie the plasticity of the brain in the face of stress and neurological insults, bridging the gap between physiological effects and the molecular and cellular events that underlie them, and has identified novel mechanisms through which stress affects the brain.

DARLENE FRANCIS
is an assistant professor in UC Berkeley's School of Public Health—Division of Community Health and Human Development. She also holds appointments in the Psychology Department and the Neuroscience Institute. Her transdisciplinary research is concerned with exploring how social experiences become biologically embedded to influence health and well-being. The study of stress is at the core of her research program.

NURTURE, NATURE, AND THE STRESS THAT IS LIFE

DANIELA KAUFER

and

DARLENE FRANCIS

People are different—from other animals and, perhaps more interestingly, from one another. One important way we differ from one another is in how we respond to stress. Why is it that when faced with the same challenges, some of us crumble, some of us survive, and some of us even thrive? How we react to stress matters; it is intimately tied not just to our vulnerability to disease states and pathologies but to our general health and well-being.

If you were to poll people at random and ask, "Where do differences in human behavior and biology come from?" most respondents, regardless of their background, education, or profession, would provide some version of this reply: "Differences between people are due both to their genes and to the environments they grew up in."

Most of us, that is, are savvy enough to know that the correct response isn't one or the other but some combination of the two. Yet scientists and laymen alike still spend too much time and effort trying to quantify the relative importance of nature and nurture. A journalist once asked the behavioral psychologist Donald Hebb which of these mattered most to personality. Hebb replied that the question was akin to asking which dimension of a rectangle was the most important, length or width?

As neurobiologists who study stress, we believe that research in this area will help reframe the study of human nature. Recent advances in our discipline make a compelling case for finally abandoning the nature-nurture debate to focus on understanding the mechanisms through which genes and environments are perpetually entwined throughout an individual's lifetime.

PHYSICAL MANIFESTATIONS OF STRESS

The idea that stress—commonly regarded as an abstract, psychological state—also affects our physical health was first proposed by the endocrinologist Hans Selye in the 1950s. Selye's work has become so well known that his concepts of "eustress" and "distress" are taught in high school and college classrooms. Subsequent advances in molecular biology and genetics have further eroded the divide between the "psychology" and the "biology" of stress. We now have the tools to observe, in exquisite detail, mechanisms induced by the brain in response to stress that cause both transient and persistent physical changes throughout the body.

Following a stressful encounter, or even just the perception of stress, the brain initiates signaling pathways that cause the release of stress hormones, such as cortisol and epinephrine (adrenaline). As systemic messengers, these hormones circulate through the body, marshaling a coordinated stress response that can affect almost every tissue and a wide range of cell types, thus contributing to a diverse range of outputs. These effects prepare the individual physiologically to deal with imminent threat. All metabolic efforts are directed toward organs and processes vital for survival (i.e., the heart and circulatory system, muscles, and brain) and shunted away from systems that are not immediately vital, such as those governing digestion and reproduction. This necessary and adaptive stress response, critical for survival, can become maladaptive when it is activated chronically, contributing to a myriad of stress-related diseases. For example, stress hormones target immune cells and alter their function, increasing susceptibility to bacterial or viral infection and cancer. They also alter the body's metabolism in ways that contribute to weight gain and diabetes, and they enact changes in the gastrointestinal tract that promote ulcers. They initiate a systemwide inflammatory response that can exacerbate aging and, together with targeted modulation of blood vessels and metabolism, precipitate major cardiovascular disease.

Given these mechanisms, it's clear that stress cannot be conceived solely in terms of subjective psychological trauma or physiological response; rather, the two are intertwined. Psychological processing of a stressor activates molecular pathways in brain and body that serve to anticipate and adapt to stress. When these pathways are induced to the extreme, or exacerbated by other challenges, they can become pathological. The effects of psychological, social, and emotional trauma can be as "real" and as consequential as diseases caused by infection, toxins, or poor nutrition. From a molecular and physiological point of view, there is not a great deal of difference between physical and psychological stressors.

MENTAL MANIFESTATIONS OF STRESS

So stress affects your physical health. But how does stress affect your mental health? Do such physical mechanisms as stress hormones play a role in mental illnesses: depression, anxiety, and other mood disorders that are frequently thought of as behavior patterns and purely mental phenomena? More particularly, how does stress affect the brain? Here one's thinking is still confounded by false dichotomies. Indeed, simply listing these binaries is a first step toward deconstructing them: mind versus body and nature versus nurture (or, in its modern version, genes versus environment). The first of these conceives of the workings of the mind as abstractions, implying that emotional states, such as stress, can exist as nonphysical ailments of the psyche. The latter imagines a linear relationship in the construction of the individual, with the genome a static blueprint creating an organism that is then reshaped throughout its life by its experience. The genome is seen as deterministic, defining the physical limits of an individual's development, and the life experience is seen as plastic. Most of us are accustomed to thinking of genes in terms of hardwiring and life experience in terms of choice, change, and possibility.

A reconsideration of the interactions between genes and the environment, however, has led to the dominant theory for how stress affects mental health. This model, termed "diathesis," states that individuals have inherent (genetic/biological) vulnerabilities that may or may not develop as disease, depending on whether the vulnerabilities are exacerbated by life experience (stress). For a real-world example of how this works, consider a hormone called corticotropin-releasing hormone, or CRH, and its receptor. This receptor is present on cells in the pituitary gland and plays a key role in initiating the stress hormone response: After a stressful experience, the brain releases CRH to activate the pituitary, which in turn initiates a hormonal cascade that ultimately circulates stress hormones throughout the body. However, not all CRH receptors are the same. In people, there are many different versions (or alleles) of the gene that encodes the CRH receptor, usually differing by a single mutation. These mutations can translate into slight differences in the expression, structure, and function of the receptor—differences that may ultimately affect the individual's vulnerability to stress-related disease. Critically, these vulnerabilities may not emerge under low-stress, or "normal," life conditions; rather, the link between a mutant CRH receptor and disease is strongest when life experience is factored in. Multiple studies have documented a link between specific alleles of the CRH receptor gene and the occurrence of depression, hyperanxiety, and suicide—along with excessive release of stress hormones in response to stress tests; however, these links are often apparent only in people and populations that have experienced high levels of stress, such as early childhood abuse. It seems that the vulnerabilities created by these alleles—that is, the way they alter the release of stress hormones—are so slight that a major taxation of the stress system is necessary to render them pathological.

This example—as well as similar ones, wherein mutations in the genes for serotonin and dopamine receptors have been shown to increase the risk of depression—reinforces the impression that the genetic contribution to pathology is fixed, creating inherent structural vulnerabilities that influence, or even dictate, an individual's response to life experience.

But can we really separate genes from environment? If we again take the molecular point of view, we may see an entirely more complex scene, in which genes and environment are not separate but constantly interacting with and affecting each other. Proteins are made from gene blueprints in the nuclei of each of the millions of cells of the body; these molecules are sent out to the cell and thence to surrounding cells throughout the body. Another contributor, from outside the body, is environmental information, which is processed by sensory systems and relayed as molecular signals sent to cells and to systems and throughout the body. Within the shared organismal landscape, molecules from both sources intermingle, interact, regulate, reorganize—doing all sorts of things. Some exert effects that are transient and malleable; others cause changes that are persistent, even irreversible. Some bind to DNA and change the way genes are expressed; others regulate sensory or hormonal systems, changing the way the environment is perceived and thereby altering the ensuing stress responses. The key questions become not what are the sources of the molecules, genes or environment, but rather what are the molecules doing? How are they exerting their effects, and what are they changing?

Recent advances in stress research are inverting implicit assumptions about gene-environment relationships. The most current data indicate that environments can be as deterministic as we once believed only genes could be and that the genome can be as malleable as we once believed only environments could be. The mind-body divide is disappearing, too, as we discover that mental phenomena have physical correlates and that understanding these physical mechanisms provides new approaches for research, teaching, and policy related to stress and health. A major challenge for the next generation of researchers will be to boldly dissolve simplistic binary abstractions, including implicit biases derived from our desire for abstraction, in order to fully embrace complex biological systems, where nuance, exception, and flexibility reign.

LONG-TERM EFFECTS OF STRESS ON THE BRAIN

There is an increasing wealth of evidence showing that environmental experiences can instantiate life trajectories in ways previously thought to occur only due to genotypes. In humans, the bulk of this data comes from population studies showing that adverse early-life stressors, both physical and emotional, are predictive of a greater incidence of mental illnesses, such as depression, anxiety disorders, and post-traumatic stress disorder. These findings are also seen in animal models, which can be used to further investigate underlying mechanisms. For example, in our own laboratories, we observe naturally occurring variations in maternal licking/grooming behavior in rats to study the effects of early-life stress on developing offspring. Much like humans, rats that receive less-than-optimal maternal care during the postnatal period often embark on developmental trajectories that persist throughout a lifetime. These trajectories often lead to adverse outcomes that include excessive responses to stressful situations (high stress-hormone release), hyperanxiety in fear-related tests, and often decreased cognitive performance (depending on the type of task).

How can even brief periods of stress at such a young age cause such persistent and long-lasting changes? What are the physical mechanisms? Even at the grossest level of observation, the effects of stress can be seen to significantly reduce the size of certain brain structures, including the hippocampus. At the cognitive level, stress may profoundly affect neural development as the brain goes through its highly plastic developmental period early in life. During this developmental time window, neurons seek out and form connections with one another, based on stimulation, forming in this way the basic neural networks that will influence cognitive patterns throughout the lifetime. Thus, stressful stimulation during this period can contribute to the formation of highly active stress-response networks with a concomitant deficit in stress inhibition, leading to permanently exaggerated fear and anxiety responses and their consequent high stress-hormone release. At the molecular level, there are a large number of proteins that modulate the stress response, many of which can be retuned in response to stressful stimuli. For example, the glucocorticoid receptor, which allows neurons to detect circulating stress hormones and initiate a negative-feedback action that terminates the stress response, is much less abundant in the brains of rats that have experienced suboptimal maternal care or early-life stress.

In addition to causing major changes during the developmental period, these types of protein modulations can allow for cumulative effects of chronic stress that build up over a lifetime. We are currently studying the effects of stress hormones on neural stem cells in the rat brain cells that are responsible for producing new neurons in adults. When these stem cells detect stress hormones, they respond by altering the expression of a variety of proteins, which in turn causes the stem cells to become glial support cells rather than new neurons. Stress is physically able to shift the fate of cells in the brain. Under chronic stress conditions, the cumulative effect of this change in stem-cell fate is likely to cause a long-term structural imbalance in the cellular composition of the brain, affecting cognition, memory, and mental health.

PLASTIC EFFECTS OF STRESS ON THE GENOME

In addition to the structural changes that stress hormones cause in brain patterning, some of the most fascinating current research is investigating the ways in which the experiences of stress are transduced into physiological signals able to directly travel to given cell nuclei to modify the genome. Environmental information contributes to plasticity in the way our genes are expressed. One way that environmental information can do this is via the up-regulation or down-regulation of proteins that bind to DNA in specific places at specific genes and modify its structure by attaching additional molecules called methyl groups to the DNA bases. This process may cause conformational and/or physical changes to DNA such that the affected sections can no longer be read by the machinery that produces proteins from the gene blueprint. For instance, recall the example of the glucocorticoid receptor, which is down-regulated in the hippocampus of rats that experience early-life stress. This reduction is caused by methylation of the region of DNA controlling expression of the glucocorticoid-receptor gene. Similarly, in the case of the fate of neural stem cells, many of the genes controlling the identity of these cells are also regulated by methylation, which effectively silences the target gene without changing its genetic code. Critically, gene methylation profiles can also be nongenomically inherited. That is, children born to a mother who is depressed may have significant vulnerabilities to stress-related mental illness and pathologies—due not to their genotypes but to changes in their gene regulation profiles.[1] Methylation is referred to as an "epigenetic" ("beyond genetics") modification. It acts like a genetic mutation, but it is not. No change is made

1 T. F. Oberlander et al., "Prenatal Exposure to Maternal Depression, Neonatal Methylation of Human Glucocorticoid Receptor Gene (NR3C1) and Infant Cortisol Stress Responses," *Epigenetics* 3 (2008), 97–106.

to the linear sequence of DNA or a given genotype; rather, the changes are evident at the level of gene expression.

Study of the stress response demonstrates that rather than being set and deterministic, our genomes can be highly responsive to life experience. Furthermore, subjective emotional experiences, which may be perceived as transitory, can induce structural changes in the organism that are persistent and perhaps irreversible. These insights come from close investigation of the actual effects of molecules and mechanisms rather than approaches that are categorical, theoretical, or abstract. We believe that these new insights have interesting implications both for future research on stress and for the implementation of social policy.

IMPLICATIONS FOR SCIENCE

When scientists conduct research, and when they train the researchers of the future, the embrace of complexity can be a daunting task. One reason that binaries such as mind-body and genes-environment are appealing is that they require you to keep track of only two concepts, two opposing or interacting forces. Simplification is convenient for students, teachers, and researchers—and especially convenient for the brain. But when dealing with complex biological systems, how do you engage students and researchers (and their brains) to track five, six, seven, and more concepts simultaneously? If you consider stress from the molecular biologist's point of view, the relevant questions are "What genes and molecules are involved? How is their activity regulated? What are their targets?" A psychologist might ask, "How does stress affect behavior and mental health?" From the standpoint of public health, the questions are "How is stress associated with disease? What are the major risk factors? What are effective interventions?"

Each discipline may promote a fairly narrow or limited set of concepts. Thomas Kuhn, in *The Structure of Scientific Revolutions,* states that a scientific community cannot practice its trade without a set of shared beliefs. Students use these fundamental tenets as the basis of their training. Kuhn argues that research is "a strenuous and devoted attempt to force nature into the conceptual boxes supplied by professional 'education.'" We believe the time has come, in our own careers as researchers and educators, to let go of outmoded and inaccurate conceptual boxes. In keeping with Kuhn, we believe that new paradigms require the reconstruction of prior assumptions and the reevaluation of prior facts. Donald Hebb was right to poke fun at the historical way many scientists have viewed differences in human nature. Yet, sixty years later, people still refuse to relinquish the conceptual "nature/nurture" boxes.

As researchers trained to study stress from opposite ends of this artificial divide, we have come to embrace the power and utility of interdisciplinary and transdisciplinary research programs and pedagogy. Across our shared research program, we are now able to tackle the topic of individual differences in stress reactivity from the most mechanistic and decontextualized level right through to multilevel/integrative concepts simultaneously: What is the molecular basis of stress? How do these processes and pathways alter the larger-scale organization of the brain and body? How do these changes affect cognition and behavior? How do these changes translate into disease, and what are the individual and social forces shaping the stress response in different populations? What are the possible interventions on the molecular (e.g., antidepressants), individual (e.g., therapy), and social (e.g., public policy) levels? Trainees in our research groups have quickly shed their previous notions about quantifying gene-environment interactions. This liberates them to fully explore how genes and environments are perpetually interacting to produce a phenotype.

IMPLICATIONS FOR POLICY

From the standpoint of public health policy, it is critically important to continually challenge the false notion that stress and its effects are purely mental states and to give those risk factors due consideration for the very real effects they have on brain and body. Though most people are accustomed to, and generally agree upon, large-scale efforts to highlight public health issues such as smoking and obesity, there is little or no effort to combat the effects that stress has on chronic disease. One example: Several studies have shown that groups subject to racial discrimination have higher mortality rates and much higher incidence of diseases that are exacerbated by stress (heart disease, diabetes, depression, and so on).[2] These differences cannot be accounted for by other socioeconomic

---

2 See, for example, D. R. Williams, "Race, Socioeconomic Status, and Health: The Added Effects of Racism and Discrimination," *Annals of the New York Academy of Sciences* 896 (1999), 173–88.

risks that correlate with racism (such as poverty), as they hold true across socioeconomic classes. Minorities in high socioeconomic classes still show a higher incidence of disease and shorter life span than their majority counterparts in the same class, despite affluence, resources, and access to high-quality health care. There is a growing consensus that these poor health outcomes are the result of the cumulative health effects of the stress of racially discriminatory encounters, as well as the perception of racial discrimination. These health effects are likely to be equally apparent in other groups suffering discrimination, including women and lesbian, gay, bisexual, and transgendered (LGBT) people. Notably, this type of stress exposure is unremitting (e.g., "feeling" like a minority or being in a low-status job). In all such cases, the health effects from chronic activation of the stress system are likely to interact with and exacerbate other risk factors (diet, drug and/or alcohol use, and so on). Furthermore, the epigenetic component of stress-related illness is likely to make disease risk particularly entrenched in groups where stress is linked to social status, since epigenetic markers and social-environmental conditions both persist across generations.

Because the early-life developmental period represents the opportunity of greatest plasticity (and hence greatest risk) in the establishment of stress-sensitive systems that will persist for life, it is the ideal target for public health interventions against stress-related disease. One advantage of this period is that developmental plasticity is as responsive to positive effects as to negative ones. Hence even modest improvement in the quality of early-life care and enrichment can have major beneficial effects on health outcomes throughout life. One inspiring example of the benefits of this approach is the Harlem Children's Zone (HCZ), an ambitious community-building project targeting all children within a hundred-block zone in Harlem. A major component of the project is aimed at early-life care, involving classes for the parents of all enrolled children in strategies for positive reinforcement and enrichment, including daily time commitments for reading to children. Graduates of this "baby college" program have remarkably improved metrics not only in educational standards (100 percent of third-graders in two of the HCZ's schools performed at or above grade level in math) but also in health outcomes, such as a lesser incidence of childhood asthma.[3] The Obama administration has pledged to establish similar "promise neighborhoods" in twenty test cities nationwide.[4] Converging data from diverse disciplines support these initiatives focused on improving early-childhood environments. A fuller understanding and identification of the mechanisms that mediate the interactions between social experiences and biologic vulnerability will allow for even better early interventions, particularly focused on children from compromised backgrounds. Programs such as the Harlem Children's Zone have the potential not just to influence the developing child but also to interrupt the generational and intergenerational cycles of adversity and disadvantage.

---

3 Visit www.hcz.org for a more complete description and summary of programs and outcomes.

4 See www.promiseneighborhoodsinstitute.org.

JON KLEINBERG

is a professor of computer science at Cornell University. His research focuses on issues at the interface of networks and information, with an emphasis on the social and information networks that underpin the Web and other online media. His work has been supported by an NSF Career Award, an ONR Young Investigator Award, a MacArthur Foundation Fellowship, a Packard Foundation Fellowship, a Sloan Foundation Fellowship, and grants from Google, Yahoo!, and the NSF. He is a member of the National Academy of Engineering and the American Academy of Arts and Sciences.

WHAT CAN HUGE DATA SETS TEACH US ABOUT SOCIETY AND OURSELVES?

JON KLEINBERG

A few months before graduating from Cornell in 1993, I attended a lecture by Carl Sagan entitled "Is There Life on Earth?" The journal *Nature* repeated the question on its cover later that year, announcing the work that Sagan was describing. Scientific inquiry often begins with a question, but on the face of it this didn't seem like one of the more difficult questions to answer.

The occasion for Sagan's Cornell lecture was a near-Earth approach in December 1990 by the *Galileo* space-craft on its way to Jupiter. As it passed toward the outer solar system, *Galileo* turned its powers of measurement on us, producing a complex array of data on the emissions coming from Earth's surface. Our own planet thus became a kind of test case for the analyses that lay ahead. In the lecture, we were asked to imagine ourselves as Martian scientists studying these observations: Was there enough in the data to recognize the signs of life, even intelligent life, here on Earth? Phrased this way, it's a kind of thought experiment that arises naturally in many settings when you're dependent on complex indirect observations in place of direct experience. Would our Martian colleagues be able to come up with the right answer?

A decade and a half later, I find myself reflecting a lot on those Martian scientists and what they could learn about us. The emissions that I imagine the Martians studying, however, are not physical or chemical but digital. Over the past fifteen years, the explosive growth of the World Wide Web has enabled a large fraction of the world's population to contribute messages, queries, links, photos, blog posts, reviews, recommendations, tweets, and other bits of text, images, and video to an ever-growing body of collective expressions. Much of it is public, even more of it is recorded, and in total it serves as an unprecedented global repository of our ideas, opinions, and communications.

The data provide new opportunities to study human knowledge and behavior on a huge scale, but doing so presents us with an enormous research challenge. Unlike scholars of an earlier time, we cannot simply "read" the entirety of what's there and draw our own conclusions. The amount of data generated by the main social media sites on the Web easily dwarfs the complete text content of the Library of Congress. As observers, we can call up and replay any particular part, but to fully understand information as a whole requires new ways of distilling it down to comprehensible dimensions.

And so we come back to the Martians. When my colleagues and I in Cornell's Computer Science Department work on computational methods to extract meaningful information from data sets of human activity, we sometimes think of the algorithms we develop as corresponding to a breed of powerful alien visitors—visitors who can observe, record, and analyze data on a scale we can't comprehend while knowing nothing of our world, our languages, or the things we care about. All they can do is notice patterns in the data as it streams by. Why so many links to a page called www.nytimes.com, many coming from sites that also link to pages called online.wsj.com and www.washingtonpost.com? Why the explosive appearance of "#MJ" on "June 25, 2009"? Why the sudden co-occurrence of the words "basketball" and "madness" every "March"? The challenge is to do more than just notice the patterns; it is to begin inferring some of the higher-level structures that generate them and ultimately to recover not just the principles we humans are familiar with but also those that have eluded our limited powers of observation and analysis.

SYNTHESIZING INFORMATION FROM THE BOTTOM UP

A year ago, we tried our own life-on-Earth experiment, using the photo-sharing site Flickr.[1] Internet users all over the world have uploaded several billion photos to Flickr.

---

1 D. Crandall et al., "Mapping the World's Photos," *Proceedings of the 18th International World Wide Web Conference*, 2009.

The site enables users to tag these photos with short (generally one-word) descriptions and geotag them with the locations where they were taken. We asked ourselves, Is there enough signal in the raw data itself to reconstruct recognizable descriptions of the significant landmarks around the world using no other sources of information?

We went about this in the following way. Imagine an Earth-sized blank sphere, on which we place a dot at the latitude and longitude of each geotagged photo on Flickr. The dots are in general sparsely scattered, but in some locations they pile up in regions of high density—places where many photos were taken. To find those high-density areas, we randomly place discs, 100 kilometers in radius, around the dot-filled sphere and computationally slide them toward where the most dots are piled. The discs come to rest at numerous local "hot spots," where there are unusually high concentrations of dots. Within each of these areas 200 kilometers across, we perform the same procedure, this time using discs only 100 meters in radius, sliding them around until they come to rest at particularly extreme, highly localized concentrations of dots. These correspond to enormous numbers of photographers cumulatively taking pictures over time at a single site 200 meters across. What we are seeing should therefore correspond to major landmarks in a broader population center.

With this in mind, we want to summarize what we've found, and again we use a computational method to let the data provide its own summary. This is a tricky proposition: even at the epicenters of these discs, the photos are awash in tags bearing simple terms such as "friends," "party," and "visit," in various languages. How can we give the name of the place itself a chance to distinguish itself against the roar of this background noise? For each place where a disc comes to rest, we choose to look not at the most abundant tag but instead at the one that most stands out from its normal pattern of usage—that is, the tag whose frequency within the disc most significantly exceeds its frequency in the world at large. This is the analogue of what a naturalist might do in a small section of rain forest, trying to find the species that most stands out in that particular area compared with the background rate at which it is normally found elsewhere.

Applying this method to our data, we get to the true test of the whole exercise. To Martians, Flickr would be just a source of raw data, but as human beings we have intuitions from our knowledge of the world that suggest what we should expect to see. So this becomes the key question: Do we find the names of cities and landmarks—our human notion of the "meaning" of those places—or are they buried by the clutter of distracting tags in the data? It turns out that the meaning spills out immediately. When we identify the densest 100-kilometer-radius discs in our data and choose the most significant tag for each disc, we get a list that begins:

newyorkcity, london, sanfrancisco, paris,…

If we pick the first of these discs and look at the most significant tag in each of the densest 100-meter-radius discs it contains, we find the tags

empirestatebuilding, timessquare, rockefeller, grandcentralstation,…

If we zoom into the second 100-kilometer-radius disc and look at the tags that stand out in its hot spots, we get:

> *trafalgarsquare, tatemodern, bigben, londoneye,...*

Even relatively far down the list, things are still comprehensible; for example, disc number 22 is named *philadelphia* and contains smaller discs named

> *libertybell, artmuseum, cityhall, logancircle,...*

and onward even to disc number 800, named *galapagos,* disc number 1,000, named *bialystok,* and beyond.

The experiment illustrates a number of principles. First, the rich intuitive meaning—all the named places and landmarks, corresponding to our human understanding of the world—was present in the data all along. It didn't require a separate source of knowledge to provide us with the names of cities and landmarks; using the right computational lens enabled us to see it directly. And because this signal is produced by a general method, it can be used on any body of information where we expect the "hot spots" to have some significance. Similar methods, when applied to a one-dimensional stream of news articles sorted by time, instead of a two-dimensional map of photos sorted by geography, can produce summaries of the leading news stories at each point in time. Applied to the one-dimensional timeline of content from a social media site, such as Twitter, related methods can produce lists of "trending topics"—lists that are updated minute by minute.

Moreover, the information has emerged in a completely bottom-up fashion; it is the synthesis of a vast number of tiny, inconsequential decisions by individual Twitter or Flickr users—a cloud of data that somehow resolves itself into a recognizable, streamlined list of cities and landmarks. This is where the view from Mars gives way to something that we, too, can understand. As in the early scenes of countless alien invasion movies, the giant discs spread out to float over all the capital cities and famous landmarks of the world. In the end, it turns out not to be as hard as you might have thought; the aliens can even figure out what we call those places.

SIGNALS FROM THE SOCIAL WORLD

Ultimately, we want to use these methods not just to discover things we expect to see, such as the names of famous locations, but to identify much more subtle effects arising from mass human behavior. There is an irony inherent in this challenge: although our powers of measurement have been increasingly able to identify phenomena taking place among vast numbers of molecules in a chemical solution, bacteria in a petri dish, or stars in a galaxy, we have been confounded by the problem of measuring and analyzing everyday social processes within human populations. It is not a mystery why this should be the case. Social life has traditionally consisted of fleeting, unrecorded interactions; to study them, we typically have had to rely on people's recollections of mundane events they were not necessarily paying attention to even as they happened.

That is what is changing. A crucial feature of online interaction is that it leaves a recorded trace, in a form that can be analyzed just as we analyzed our massive Flickr data set. Vast digital trails of social interaction allow us to begin investigating questions that have been the subject of theoretical inquiry and small-scale analysis for a century or more—a reflection of the broader principle that science moves forward whenever we can take something that was once invisible and make it visible.

One focal point for recent research on these data sets has been the investigation of social influence: the tendency of people's beliefs, opinions, and behaviors to change over time to more closely reflect those of their friends. Social influence is an ideal case study for these large-scale techniques, since it is a ubiquitous, hard-to-measure aspect of social life and one that can be explored using digital data to track changing patterns of behavior across networks of social interactions. Using digital records, we can see to what extent people's decisions to join online groups, contribute to tasks in large online projects, express opinions in online review and discussion venues, or adopt new features in social media sites depend on whether their friends have done so.

This process has often been referred to as "social contagion," since it spreads from person to person through a social network, much as biological contagion does. But there are significant differences between social and biological contagion. Adopting a new behavior, in a social context, involves decision making; instead of simply "catching" a behavior from a friend, people generally use complex rules, at least implicitly, in assessing how to behave given their social context. It is the architecture of these rules that we can try to infer from the data.

CAUSALITY AND PREDICTABILITY

We encounter two particularly intriguing challenges as we delve into the question of social influence: causality and predictability. Friends are generally similar to one another in many dimensions already—this similarity is often what drew them together in the first place—so when you see two friends engaging in a shared behavior, it could simply be the result of their underlying similarity rather than any overt influence. The direction of cause and effect thus becomes hard to determine: How can we gauge the extent to which the similarity caused the friendship and the extent to which the friendship caused the similarity? Once you appreciate this circularity, the whole metaphor of contagion begins to feel a bit precarious. My Cornell colleague the sociologist Michael Macy likens the issue to the distinction between flu and hay fever: in both instances, we see a pattern of symptoms spreading through a population, but in the former case it represents genuine contagion, whereas in the latter case it simply represents a shared response to similar underlying conditions.

This question of causality is one for which online data can help separate the two different effects at work. For example, recent research on the adoption of new features in an instant messaging system involved a population of millions of users with rich individual profiles; it thus became possible to carefully quantify and control for the users' similarities to one another and then to estimate how much of the observed adoption of features might be attributable to forces of influence operating beyond this baseline similarity.[2] In a different study, on Wikipedia, we analyzed the exact time at which each edit to an article and each communication between two editors took place, and we used this detailed information to investigate how much of the similarity in the two editors' behaviors pre-dated their first social interaction and how much of it arose after that.[3] Consistently, these studies suggest a nuanced picture in which similarity and influence coexist; much of what we might initially think of as influence is in fact explainable by the underlying backdrop of similarity among friends, although evidence in support of influence remains.

Studies at this new level of resolution raise the hope that we might eventually be able to use these findings for predictive purposes, applying models of social influence to predict which songs, movies, or books will become hits, which political candidates will attract widespread support, or even, over longer time scales, which urban locations will become the kind of iconic landmarks found in our Flickr data. But recent work by Matt Salganik, Peter Dodds, and Duncan Watts of Columbia University offers a cautionary note, arguing that the outcomes of social influence over

2 S. Aral, L. Muchnik, and A. Sundararajan, "Distinguishing Influence-Based Contagion from Homophily-Driven Diffusion in Dynamic Networks," *Proceedings of the National Academy of Sciences,* 106 (2009), 21544–49.

3 D. Crandall et al., "Feedback Effects Between Similarity and Social Influence in On-line Communities," *Proceedings of the 14th ACM SIGKDD International Conference on Knowledge Discovery and Data Mining,* 2008.

time might, at least in part, be inherently unpredictable.[4] They conceived of the following intriguing thought experiment: If we could roll history back twenty years and run it forward again, would the same creative works and celebrities appear? Would, say, the Harry Potter books again sell hundreds of millions of copies or would new mega-bestsellers emerge?

For obvious reasons, this is not an experiment you can run in the real world. But in the online domain you can create a close analogue, and this is what Salganik, Dodds, and Watts did. They created a music download site with songs by obscure artists—real music groups, but not people you were likely to have heard of before coming to the site. Users could listen to songs and download the ones they liked, and a prominent and continually updated "leader board" listed the songs with the most downloads. As expected, the leader board tended to sway people toward the songs that were already doing the best, causing a "rich-get-richer" effect. But here was the twist: the researchers were actually running eight simultaneous copies of the site, and each user was randomly deposited into one of these "parallel universes." The different universes were all operating under exactly the same conditions—nothing was deliberately being altered between one copy and another—but they were allowed to evolve independently. As a result, different songs leapt to early high positions on the leader boards in different universes, and the songs with the early

---

4 M. Salganik, P. Dodds, and D. Watts, "Experimental Study of Inequality and Unpredictability in an Artificial Cultural Market," *Science* 311 (2006), 854–6.

advantages tended to remain out in front as time went on. A ninth universe served as a control—there was no leader board or any other form of social feedback in this one, so the success of songs here was presumably based primarily on how many listeners actually liked them. Compared to this baseline, the eight other universes displayed a surprising level of variability, although the songs judged best in universe number nine never ended up at the bottom of any of the other eight and the songs judged worst in number nine never ended up at the top anywhere else. The implication is clear: The winning choices seem to have an important "accidental" component. A song can acquire an early lead for reasons that seem inconsequential, based on the tastes of the people who sample it first, but once this lead is established—for whatever reason—it has a good chance of being perpetuated through social feedback. And through this online construction of mirrored worlds, we can begin to think more carefully about how similar effects may be at work in the real one.

THE MERGING OF ONLINE AND OFFLINE

It is fair to ask how much the effects we're seeing are a consequence of the digital nature of online interaction and how much they apply to offline social life as well. For instance, despite the enormous amount of data marshaled to perform our initial studies of photos, it was still restricted to the collection of photos uploaded to Flickr and thus reflects only a particular slice of the photo-taking population. This likely helps to explain features such as the appearance of "sanfrancisco" as the third densest population center.

There is no conclusive answer to this question of online versus offline, although several potential answers are the subject of current investigation. One approach suggests that online interaction, despite its characteristic trappings, is still governed by many of the same social forces shaping our offline interactions. Recent studies of Facebook, for example, suggest that although users accumulate and display huge lists of "friends," the friendships involving regular communication present a pattern that is sparser and more consistent with what we see in the offline world. Moreover, Facebook friendships tend to occur at relatively close-range geographic scales, despite the Internet's indifference to geography.[5] But that doesn't mean that the daily rhythms of social interaction are remaining unchanged; the broader point is that our online and offline forms of interaction are steadily merging, and this tendency will only increase as we welcome a new generation that, justifiably, has a hard time identifying where the technology ends and their "ordinary lives" begin. I try to imagine what early typewriter salesmen would have thought if I could have predicted for them that someday I would be sitting on a bus next to four teenage friends ignoring one another as they furiously typed gossip to distant counterparts on miniature QWERTY keyboards. The sociality of

---

5 L. Backstrom, E. Sun, and C. Marlow, "Find Me If You Can: Improving Geographical Prediction with Social and Spatial Proximity," *Proceedings of the 19th International World Wide Web Conference,* 2010. See also D. Liben-Nowell et al., "Geographic Routing in Social Networks," *Proceedings of the National Academy of Sciences* 102 (2005), 11623–8, and J. Kleinberg, "Complex Networks and Decentralized Search Algorithms," *Proceedings of the International Congress of Mathematicians,* 2006.

the digital world strains against the sociality of the physical world, striving for a new equilibrium in which the two are no longer really separate.

This is the trajectory that our modern information systems have taken. As they acquired enormous audiences that were linked through social interaction, their fundamental design constraints came to involve not just computational efficiency, network bandwidth, and processing power but also a range of concerns that are inherently rooted in collective human behavior, including attention, participation, similarity, and influence. And as these systems continue to accumulate digital trails over time, we should not think of them as recording only an aggregate picture; disconcertingly accurate portraits of each of us as individuals begin to come into focus. This raises strong privacy concerns and leads to active research on methods for the preservation of online privacy.

But it also leads to something else, a final reflection as we look forward: the prospect of software that compiles such detailed statistical profiles of your behavior that it ultimately comes to know more about you than you do. Powerful digital assistants that can learn from every message you've sent, every purchase you've made, and every song you've listened to can have many uses—think of the possibilities for personal organization and appropriate recommendations—but even if they reserve their advice just for you, will you always want to hear what they have to say?

I periodically bump into this question, often inadvertently, through my work, and it is always more jarring than I expect. A few years ago, I was testing some of the ideas that later went into our Flickr study, using a decade's worth of my own archived e-mail. I was building timelines of this e-mail collection by finding words that underwent sharp bursts of intensity and then faded again—a kind of "trending topics" for my own life. To my surprise (I should have seen it coming), the early words in the first part of the timeline were filled with the names of people with whom I had once been in close communication and had since lost contact. They were people I had meant to keep in touch with—in fact, had thought I was still in touch with—but of course it was hard to argue with what the timeline was telling me. It was an insight into changes in my life that had been too gradual for me to notice and that were disturbing when brought into sharp relief. It's not the kind of thing you enjoy having pointed out and not the kind of judgment you expect to be rendered by a few hundred lines of computer code—dispassionate, alien, and working purely from the data.

Sometimes the view from Mars can be painful.

COREN APICELLA
earned an MS in evolutionary psychology at the University of Liverpool in 2001 and a PhD in biological anthropology at Harvard University in 2009. She is now a postdoctoral research fellow at Harvard Medical School in the Department of Health Care Policy. She has published in the fields of evolutionary psychology, behavioral economics, and behavioral genetics. Her work has been featured in media outlets worldwide, including CNN, Nightline, BBC, *and* The New York Times.

ON THE UNIVERSALITY OF ATTRACTIVENESS

COREN APICELLA

So powerful and pervasive are the forces that attractiveness exerts on our lives that scholars have long considered what makes people attractive and why. Beauty's ability to influence, inspire, and intoxicate is proverbial, part of the human condition. The profound nature of its effect was perhaps first recorded in Greek mythology, with the Trojan War, credited to Helen of Sparta, the beautiful daughter of Zeus, later famously celebrated by Christopher Marlowe as "the face that launched a thousand ships." The pursuit of beauty is ubiquitous—witness the facial scarification practiced by African tribes, the use of the poisonous belladonna plant for pupil dilation in ancient Italy and Spain, the vast array of cosmetic procedures currently performed in the West. Clearly, individuals will go to great lengths to improve their appearance.

Science has recently called into question the old adage that beauty is in the eye of the beholder. What makes a person attractive may not be solely a matter of individual taste or a particular culture but instead may reflect evolved preferences—predilections handed down from our prehistoric ancestors. Isolated and evolutionarily relevant groups such as hunter-gatherers may hold the key to important questions underlying the origins of attractiveness.

This evolutionary account assumes that individuals who preferred mates with traits signaling superior physical qualities such as health and fecundity would have had greater reproductive success, or what evolutionary biologists call fitness, than individuals without these preferences. Whether attractiveness preferences are indeed adaptations to solve the problem of mate choice is an open empirical question. One way to test this theory is to examine whether preferences for certain traits are universal and evolutionarily significant.

My quest to understand the natural origins of attractiveness preferences led me to the African savannah near Lake Eyasi in Tanzania. This is a mostly dry lake in the Great Rift Valley, just south of the archaeological site of Laetoli, where Mary Leakey, in 1976, discovered a line of 3.6-million-year-old hominid footprints preserved in volcanic ash. Among these remnants of our past reside the Hadza people, one of the last remaining hunter-gatherer populations in Africa. As such, they may hold the key to important questions about our humanity.

The study of hunter-gatherers is appealing not only because they provide the most extreme departure from modern life but also because their way of life is evolutionarily relevant. We have been hunter-gatherers for most of our species' time on the planet, practicing agriculture for only 5 percent of the 200,000 years *Homo sapiens* has been in existence. Until about A.D. 1500, a third of the planet was still inhabited by hunter-gatherers. Selecting an optimal mate is not an exclusively modern problem, nor is it a uniquely human problem; animals discriminate in their choice of mates on the basis of attractiveness. Thus there is ample reason to assume that our hunter-gatherer ancestors also exhibited preferences for mate choice. Since the Hadza live under conditions that approximate those in which our species spent most of its history and likely face many of the same problems our ancestors faced, they may provide clues as to what traits our ancestors found attractive and why.

The Hadza are an ideal population to study attractiveness for a number of other reasons: First, they are relatively isolated from Western media and ideals. Attractiveness research has been criticized for its overreliance on Western college populations, which limits its ability to eliminate cultural-construct claims for human beauty preferences. If attractiveness preferences are adaptations to enhance reproductive success, we should expect to observe some universal agreement on what traits are considered attractive. The Hadza people are among the best population of foragers to test for universality. While they have had some exposure to the West—the number of ethnotourists who visit them has increased in the past few years—Hadza culture and tradition have been highly conserved, as can be inferred from reading ethnographic descriptions made a century ago.

In investigating whether preferences for attractiveness are evolutionary adaptations, it is important to examine whether features deemed attractive are related to health and/or reproductive success. Assessing this in modern societies is problematic because of advances in medical and cosmetic technologies and the widespread use of contraceptives. The Hadza do not use contraceptives, and their access to medical care is highly limited. Moreover, they endure a number of environmental stressors that we in the West are not exposed to, the most obvious being malaria and intestinal parasites. The Hadza sleep on the ground, either outside or in grass huts. Their subsistence lifestyle of hunting and foraging is energetically demanding and intensive. Infant mortality is extremely high, and the overall life expectancy at birth is thirty-three years. All these factors create an ideal testing ground for examining relationships between attractiveness and health and reproduction.

Like people in the West, the Hadza value physical attractiveness in a mate. The Florida State University anthropologist Frank Marlowe, who first introduced me to the Hadza, asked Hadza men and women to list the traits most important to them in choosing a spouse. Attractiveness was an important criterion, surpassed only by good character and foraging ability. But I was curious about whether the Hadza find the same things attractive as Western people do. I decided to focus my research on facial, body, and vocal traits found attractive in the West.

There is no one trait that constitutes beauty; rather, several components work together to create an aesthetically pleasing face, body, or voice. Traits that affect facial attractiveness include symmetry, averageness, sexual dimorphism, cleanliness, and youthfulness. Components of body attractiveness may include body mass index, waist-to-hip ratio (WHR) in women, waist-to-shoulder ratio (WSR) in men, and muscularity. Voice attractiveness may be determined by intonation, peak frequency, and fundamental frequency (pitch). Here, I will focus on one aspect each of the face, body, and voice that have been found attractive in the Western population: facial "averageness," WHR in women, and vocal pitch.

AVERAGENESS

Average faces—that is, faces possessing traits common in a population—are generally perceived as attractive. Anthropological and psychological research has demonstrated that composite images made by superimposing faces are judged as more attractive than almost all individual faces. Individual faces are perceived as more attractive when their configurations are manipulated to be closer to average configurations for their sex. Researchers have suggested that the preference for averageness is based on associations with reproductive success, since natural selection would have eliminated deleterious mutations that move traits away from the mean. That is, traits that are common in a population are common because they worked.

I decided to investigate whether the Hadza also prefer average faces. If they do, this would provide strong evidence that the preference is biologically based and possibly an adaptation for choosing healthy and fecund mates. To test preferences for averageness, my collaborator Tony Little and I generated pairs of computer-morphed more-average and less-average faces for each sex. We made two sets of these contrasting stimuli, the first based on a collection of Caucasian faces and the second based on a collection of Hadza faces. I then asked both Hadza and European subjects to choose the opposite-sex face they found most attractive from each pair. The Westerners preferred the more average faces when judging both Caucasian and Hadza faces, whereas the Hadza subjects found averageness more attractive in Hadza faces and ignored it in Caucasian faces.

The fact that the Hadza prefer averageness in Hadza faces but not in Caucasian faces is unsurprising; the different visual experiences of the two populations likely explain the difference in preference. The Hadza have had very little exposure to Caucasian faces apart from us researchers and the occasional tourist who might visit them. Therefore, it is unlikely that they have had enough visual experience to form a fully applicable mental representation of what an average Caucasian face might look like. Westerners, for their part, have had plenty of visual experience of both Caucasian and African faces. Though the study strongly suggests that the preference for averageness is biologically based, it also demonstrates the importance of experience. It is experience that influences our mental representations of what is average and hence what we find attractive in the opposite sex. In future work with the Hadza, I plan to examine whether increased exposure to Caucasians generates a preference for averageness in their faces.

FEMALE BODY SHAPE

Universality is taken to mean that there are some regularities in preferences that are common across cultures.

Universality in no way implies, however, that preferences are invariant with local conditions; on the contrary, the fact that many preferences vary across cultures suggests a certain amount of plasticity. Preferences may have been shaped by selection to be conditional on local circumstances with the value of attraction to specific traits being modulated by specific environmental inputs. For instance, preference for fatness may vary adaptively, depending on access to resources within a society.[1]

In previous work with the Hadza, it was shown that Hadza men prefer women with a high WHR. This finding contrasts to a large body of literature finding that men in the West prefer women with a low WHR. The Hadza may prefer a high WHR because it is associated with greater fat storage. In the Hadza, fat reserves are likely to be highly valued, since food is sometimes scarce. Nevertheless, abdominal fat is generally considered unhealthy; it has been associated with a number of metabolic diseases and reproductive cancers. Graznya Jasienska, a biological anthropologist who was trained in the same laboratory at Harvard that I was, has found that in Polish women a lower WHR is associated with a higher estrogen-to-androgen ratio and possibly higher fecundity.

Studies reporting relationships between health and WHR obtain a woman's WHR by dividing her waist measurement by her hip measurement, which is taken at the

---

1 P. J. Brown and M. Konner, "An Anthropological Perspective on Obesity," *Annals of the New York Academy of Science* 499 (1987), 29–46; J. L. Anderson et al., "Was the Duchess of Windsor Right?: A Cross-Cultural Review of the Socioecology of Ideals of Female Body Shape," *Ethology and Sociobiology* 13 (1992), 197–227.

widest part of the hips, where the buttocks protrude the most. However, preference studies have been conducted using two-dimensional images in which women are shown in frontal view, ignoring the role of the buttocks. If the relative contribution of the hips and buttocks to the actual WHR did not vary among women across continents, frontal views would suffice, but they will not if some women have wider hips and others more protruding buttocks. Nor will frontal pictures tell us the full story of men's preferences for women's WHR. Therefore, Frank Marlowe and I decided to study preferences for WHR in profile view. We showed pictures of women with varying WHRs in profile view to both Hadza and American men. We found that Hadza men preferred a lower WHR in profile view. In profile view, a lower WHR signifies more protruding buttocks. Since Hadza men prefer a higher frontal WHR but a lower profile WHR, and since both contribute to the actual WHR of women, our results imply that there is less disparity between American and Hadza preferences for the actual WHR of real women than was previously indicated by studies that used only frontal pictures. However, Hadza men do still prefer women with a larger waist and more protruding buttocks than do American men, likely reflecting the importance of fat storage to the Hadza.

VOICE PITCH

Vocal attractiveness, like physical attractiveness, may signal mate quality. Voice pitch, a sexually dimorphic trait (that is, a trait that differs in men and women), is the perceptual correlate of vocal fundamental frequency. Most studies in the West have found that women find lower-pitched men's voices attractive. Lower-pitched voices suggest that their owners are dominant, older, healthier, more masculine. Correspondingly, men find higher-pitched voices in women to be more attractive and perceive their owners to be more feminine, healthier, and youthful.

Voice pitch in men, because of its association with testosterone (the more testosterone, the lower the pitch), may be a sign of genetic quality: Production of a low-pitched voice is thought to be costly, since testosterone is an immunosuppressant; to produce a low-pitched voice, you must be of sufficient genetic quality to withstand testosterone's deleterious effects. In women, voice pitch may indicate reproductive ability, signaling a low androgen-to-estrogen ratio. Researchers have found that voice pitch increases as women near ovulation.[2]

With the help of David Feinberg of McMaster University, an expert on voice attractiveness, I tested preferences for voice pitch in the Hadza. I was specifically interested in whether voice pitch affects assumptions that individuals make about a male speaker's ability to hunt and a female speaker's ability to gather. I had the Hadza listen to voices of both sexes whose pitch had been manipulated. We found that our subjects associated lower-pitched voices in men and women with greater perceived hunting and gathering abilities, respectively. Both hunting and gathering are strenuous and labor intensive, requiring a good deal of strength and muscularity. Therefore, the perceptual link

2 G. A. Bryant and M. G. Haselton, "Vocal Cues of Ovulation in Human Females," *Biology Letters* 5 (2009), 12–15.

between better hunting and gathering and voice pitch may reflect testosterone's joint effect on voice pitch and muscle strength (also associated with a greater testosterone level).

We also tested Hadza voice-pitch preferences in choosing a marriage partner. Although Hadza men value gathering ability in women, they also place considerable importance on youth and reproductive potential. In line with this preference, Hadza men typically chose women with higher-pitched voices as marriage partners. Marrying the best gatherer likely does not increase a man's reproductive success as much as marrying a woman with the greatest reproductive potential.

Although the Hadza women we tested perceived men with lower-pitched voices to be better hunters, they did not display a preference for voice pitch when choosing marriage partners. Since approximately half our sample consisted of women who were breast-feeding, we tested whether this lack of preference could be explained by variation in women's preferences according to their reproductive state. We found that women who are breast-feeding preferred men with higher-pitched voices for marriage partners, whereas women not breast-feeding preferred men with lower-pitched voices. This finding fits nicely with a number of studies that suggest that women display strategic shifts in masculinity preferences depending on the balance between choosing men who exhibit high testosterone levels (and thus good genes) but may be less likely to invest in relationships and offspring and men who have less testosterone but may make faithful husbands and fathers.

Although studies have found that low-pitched male voices are generally preferred by women and high-pitched female voices by men, likely because these are signs of superior mate quality, the validity of these evolutionary explanations hinges on whether these traits are indeed associated with greater reproductive success. We decided to test whether Hadza women with high-pitched voices and Hadza men with low-pitched voices had more children. To do this, in each Hadza camp I visited, I recorded all adult men and women speaking the Swahili word "Hujambo" (which loosely translates as "Hello"). I also noted their reproductive histories.

My results indicated that whereas voice pitch was unrelated to reproductive outcome in women, men with low voice pitch reported having fathered more children. Hadza men can increase their reproductive success by reproducing at an earlier age, having multiple mates, or engaging in serial monogamy with younger partners. Although high-pitched voices in women may signal reproductive potential, we did not find it to be related to reproductive outcomes. This is not particularly surprising, since there was less variance in reproductive outcomes in our sample of women than in our sample of men. If there is a relationship between voice pitch and reproduction in Hadza women, the relationship may be subtle and not observable in a small sample.

CONCLUSION

The research presented here was part of a larger effort to understand human preferences for attractiveness from an evolutionary perspective. For a long time, the unstated assumption in most writing on physical attractiveness was that standards of beauty are arbitrary and culturally determined. In recent years, this view has been called into question as scholars, drawing on evolutionary theory, have demonstrated a surprising degree of cross-cultural consistency in attractiveness preferences.

Evolutionary theory provides a strong framework for examining attractiveness preferences. Such a framework predicts that traits signaling fitness benefits should be universally preferred. With few exceptions, studies on attractiveness have drawn either on culturally homogeneous populations (e.g., American college students) or populations that have had significant exposure to Western culture. Therefore any agreement found in attractiveness preferences within or among cultures may arise due to common exposure to Western standards of beauty, as opposed to a shared evolved psychology for appraising attractiveness. The primary contribution of my work with the Hadza is that it establishes that many of the regularities uncovered about attractiveness judgments across cultures also appear to hold true in an evolutionarily relevant population. Since the usual criticisms of cross-cultural testing for universality apply with much less force to this population, the results reported here offer the strongest evidence to date in favor of human universals for attractiveness preferences.

LAURIE R. SANTOS
is an associate professor of psychology at Yale University and the director of its Comparative Cognition Laboratory. She received her BA (1997) in psychology and biology and her PhD (2003) in psychology from Harvard University. She has investigated a number of topics in comparative cognition, including the evolutionary origins of irrational decision making and prosocial behavior. She is the recipient of Harvard's Goethals Award for Teaching Excellence, Yale's Greer Memorial Prize for Outstanding Junior Faculty, and the Stanton Prize from the Society for Philosophy and Psychology for outstanding contributions to interdisciplinary research.

TO ERR IS PRIMATE

LAURIE R. SANTOS

It was the final shot of the tournament for the world's number one player. After three tense rounds in the 2009 Barclays Tournament, Tiger Woods was now one putt away from another tournament win. His fairway shot was nearly perfect—his ball had landed just seven feet from the hole. Making this putt would earn him a birdie on the last hole and a hefty payoff. He practically beamed as he stepped up to a putt he had sunk thousands of times before. After the tournament, he would be asked if he had approached this particular shot any differently. "Absolutely not," he would emphasize. "Every putt you hit is the same process. Go up there. Be committed to what you're going to do. Hopefully it goes in." Only this time it didn't. A stunned crowd watched in disbelief as the ball skimmed past the hole. Tiger's shot was just a bit off, but it cost him the lead. He took another putt, made par, and lost nearly a million dollars in winnings.

For professional golfers, every putt is a risky decision, one that can have big financial consequences. A putt is a reasonably simple goal-directed action, yet each stroke requires more than just motor skill. Good golfers sink putts because they're also good decision makers. Every putt requires a host of tough choices. Besides having to estimate how the ball will break, a player must choose between playing it safe—going with a softer stroke that will mean an easier following shot if things go badly—or going for the hole at the risk of overshooting. As the above example illustrates, even the best golfer in the world can make errors.

Those of us who aren't golfers are not immune to the difficulty of making risky decisions. Though we (usually) play for smaller stakes than Tiger, we too spend our days navigating risky choices that can have significant consequences for our health, bank account, and overall well-being. The question of how to best make decisions has fascinated humankind for centuries. For economists, the answer has always been relatively simple: making good decisions is a simple act of comparison shopping. A smart decision maker should start by listing all the possible choices for a given decision and then estimate the average payoff of each individual choice. Once the decision maker has all this information handy, he just needs to pick the choice with the highest expected payoff. Simple, right? Unfortunately, in practice the strategy of maximizing your expected return runs into a number of thorny issues.

First, most decisions don't come with a finite set of nicely lined-up choices. For our biggest decisions in life—finding a mate, choosing a career, and so on—it's often hard to know exactly how many options are at our disposal. In addition, we often have limited information about how the various choices we can identify will actually affect our happiness. For all these reasons, real decision making usually fails to live up to economists' lofty standards. Given the difficulty of maximizing payoffs, it's no surprise that we make tons of bad mistakes all the time. What is surprising, is that we don't just make random mistakes, we seem to make systematic mistakes. We don't just experience a catastrophic cognitive meltdown when facing hard choices; we instead systematically switch on a set of simple (though mostly irrational) strategies to weigh those choices.

To witness one of your own irrational strategies in action, consider the following scenario: Imagine that you are an economic adviser to the president of the United States. Your goal is to choose a course of action that will reduce the rate of housing foreclosures for the 3 million home owners currently in danger. Two plans are on the table. If Plan A is implemented, the government will be able to save 1 million homes. If Plan B is implemented, there's a one-in-three chance that the government will be able to save 3 million homes and a two-in-three chance that no homes will be saved. What's your advice?

You probably suggested that the president go with Plan A. Any plan guaranteeing that at least some people will keep their homes seems like the better option. Fair enough. But what if the options are slightly different? Imagine a choice between two new plans, C and D. If Plan C is implemented, 2 million people will lose their homes for sure. If Plan D is adopted, there's a two-in-three chance that 3 million people will lose their homes and a one-in-three chance that no one will lose his home. Here you might advise the president to go with Plan D. It's a riskier option, but it also offers the possibility that no homes will be lost. When the psychologists Daniel Kahneman and Amos Tversky tested undergraduates using similar scenarios, most of their subjects showed the same pattern: they preferred Plan A to Plan B and Plan D to Plan C.[1] The problem with this pattern of decision making is that the two sets of plans are identical. Plans A and C are statistically indistinguishable (since the 3 million homes are at stake in all scenarios, a result in which 1 million people keep their homes is identical to a result in which 2 million people lose theirs). The same is true of Plans B and D. As Kahneman and Tversky observed, small changes in the wording of a problem have a big effect on our preferences. When plans are presented in terms of the number of houses, lives, or dollars saved, people tend to play it safe, but when plans get us thinking in terms of houses, lives, or dollars lost, we switch to riskier tactics.

Why does a simple change in wording so critically influence our decisions? Kahneman and Tversky discovered that the culprits are two psychological biases: reference dependence and loss aversion. The first of these is our tendency to see things not in absolute terms but relative to some status quo. Most people think about their decisions

1 D. Kahneman and A. Tversky, "The Framing of Decision and the Psychology of Choice," *Science* 211 (1981), 453–58.

not in terms of their overall happiness or total net worth but as gains or losses relative to some reference point, usually the here and now. A $20 parking ticket won't have a significant effect on our life savings, but it's still a negative change from our current wealth level, and thus we tend to find the event salient. The parking-ticket example also highlights the second psychological bias at work: loss aversion. We generally avoid situations in which we could incur a loss. Indeed, Kahneman and Tversky's studies have shown that we work twice as hard to prevent being in the red as we do to seek out opportunities to land in the black.

Reference dependence and loss aversion appear to wreak havoc in a number of real-world situations. Investors tend to view the value of a stock not in absolute terms but relative to a salient reference point: what they paid when they bought it. Averse to the loss of selling below the purchase price, many investors irrationally hold on to stocks while they're dropping in value.[2] These biases also cause problems in the housing market; people are averse to selling for less than what they paid, which has led some families to decline such offers.[3] Indeed, these biases are so widespread that they affect the scores of professional golfers. A golfer's only true measure of success is his or her final score, but each hole has a salient reference point: par. The economists Devin Pope and Maurice Schweitzer analyzed more than 1.6 million PGA Tour putts to deter-

2 T. Odean, "Are Investors Reluctant to Realize Their Losses?," *Journal of Finance* 5 (1998), 1775–98.

3 D. Genesove and C. Mayer, "Loss Aversion and Seller Behavior: Evidence from the Housing Market," *Quarterly Journal of Economics* 116 (2001), 1233–60.

mine whether players tended to perform differently when putting for birdie and eagle (i.e., strokes that put them under par) than when faced with comparable putts for par and bogey (i.e., strokes that could put them over par). Consistent with loss aversion, players were more accurate when putting for par and bogey, meticulous in their attempt to minimize the "loss" of going one or two strokes over par. Players putting for birdie or eagle, in contrast, were about 2 percent more likely to miss the hole. This small percentage of errors adds up fast—just ask Tiger Woods. Tiger's loss aversion statistic was one of the highest on the tour; that is, he was 3.6 percent more likely to miss his birdie putts than his par putts. Indeed, this bias may have been what cost him the 2009 Barclays on the eighteenth hole.[4]

Why do house sellers, professional golfers, experienced investors, and the rest of us succumb to strategies that make us systematically go wrong? A few years ago, my Yale colleagues Venkat Lakshminarayanan and Keith Chen and I decided to try to get to the bottom of this question. After reviewing examples in which people succumb to these biases time and again, we started thinking that reference dependence and loss aversion might be more fundamental than economists had previously thought. This led us to a somewhat radical idea: perhaps these biases are a natural part of the way we view our choices, a result of a long evolutionary legacy. If so, we hypothesized, humans might not be the only species to use these poor decision-

---

4 D. G. Pope and M. Schweitzer, "Is Tiger Woods Loss Averse? Persistent Bias in the Face of Experience, Competition, and High Stakes" (2009), http://ssrn.com/abstract=1419027.

making strategies. Rather than investigating the biases of human subjects, we decided to test whether similar errors showed up in the decision making of one of our primate relatives: the capuchin monkey, whose last common ancestor with humans lived around 35 million years ago.

Our question was whether capuchins would show humanlike patterns of reference dependence and loss aversion, even though they lacked experience with the kinds of economic problems that typically lead human decision makers astray. Our first challenge was figuring out how to demonstrate loss aversion and reference dependence in monkeys. Capuchins aren't all that good at investing in stocks or playing golf, so the way ahead was unclear. In the end, we decided to give the monkeys some money and see whether they could be taught to use it.[5]

The idea of teaching monkeys to use money might seem daunting, but the process took only a few months. We began by introducing them to a token economy. The capuchin "tokens" were coin-sized metal discs that could be traded with experimenters for food. Although the monkeys didn't know what to do with the tokens at first, within weeks they were handing tokens to experimenters and holding out their hands for the food. We then allowed the capuchins to use the tokens in a real economy. Each monkey was given a wallet of tokens before entering a "market," in which two "salesmen"—research assistants—offered it two different kinds of food at two

---

5 M. K. Chen, V. Lakshminarayanan, and L. R. Santos, "The Evolution of Our Preferences: Evidence from Capuchin Monkey Trading Behavior," *Journal of Political Economy* 114 (2006), 517–37.

different prices. The monkeys could spend their tokens to buy whichever treat they wanted. Like human shoppers, our monkeys quickly became skilled at maximizing their token value. They bought more food from experimenters who gave them a better deal. They bought more food during "sales," when prices were cheaper. They carefully weighed the risks of dealing with unreliable salesmen who switched their behavior over time. Our monkeys' performance so closely mirrored that of human consumers that the data fit perfectly with formal economic models of human market choice.

We were now ready to ask the real question of interest: Would monkeys' market behavior be affected by loss aversion and reference dependence? To study this, we set up two situations: in both, the salesmen didn't always hand over the number of apple pieces they had originally displayed—sometimes the monkeys got more pieces than they had been offered, sometimes fewer. We hypothesized that the monkeys might make their choices based not just on how many apple pieces they managed to get overall but also on how many they got relative to how many they originally saw displayed. In other words, we predicted that the monkeys would use the original offer as a reference point and, like professional golfers thinking about par or birdie, make their decision based on whether the payoff seemed like a loss or a gain relative to that reference point.

First, each of the monkeys got to choose between dealing with Salesman A or Salesman B. Salesman A would always offer a monkey one piece of apple and, when the monkey made its payment of a token, add a second piece as a bonus. Salesman B was more of a risk. He also began by showing a monkey one piece of apple, but his apple payoff changed across trials: on some trials, after receiving the monkey's token, he added a large bonus of two apple pieces, while on other trials he gave no bonus at all. Just like humans asked to choose between the government plans A and B, our monkeys bought more food from Salesman A than from Salesman B. Like people, they preferred to play it safe when dealing with bonuses.

We then introduced monkeys to two new salesmen, C and D, whose payoffs felt like losses. Salesman C always gave a small but consistent loss—he showed the monkeys three apple pieces but gave them only two in return for the token. Salesman D was riskier. He began by showing the monkeys three pieces and sometimes gave them all three but other times gave them only one. As predicted, our monkeys took greater risks when their token payments felt like losses; that is, they consistently preferred to trade with risky Salesman D over reliable (but shortchanging) Salesman C.

Overall, our monkeys behaved just like humans tested in Kahneman and Tversky's scenarios. They thought about the market in terms of arbitrary reference points and responded to payoffs differently depending on whether the payoffs appeared to be gains or losses relative to those reference points. In this and other studies, monkeys seemed not to consider their choices in absolute terms. Moreover, they made decisions differently when dealing with losses than when dealing with gains. These findings suggest that the biases that human decision makers show may be far more fundamental than originally thought. The biased strategies that cost Tiger Woods millions of dollars each year may be at least 35 million years old.

The discovery that loss aversion and reference dependence may be deeply evolved psychological tendencies has important implications for our ability to overcome these biases. For years, economists have assumed that decision makers would stop using irrational strategies in the face of enough negative financial feedback. Unfortunately, there is growing evidence that people don't drop these strategies as soon as they become costly. Pope and Schweitzer estimate that loss aversion causes even experienced professional golfers to lose more than a million dollars a year, yet nearly all golfers on the tour exhibit these biases. Similarly, investors tend to hold on to losing stocks even after suffering repeated losses because of doing so. Our capuchin findings suggest an answer to why these biases might be so hard to overcome: reference dependence and loss aversion may be as deeply ingrained as some of our other evolved cognitive tendencies. Just consider how difficult it is to switch off our natural fondness for cheesecake, our squeamishness about bugs, our disgust at a pile of feces. When natural selection builds in a strategy, it's hard to get rid of. If reference dependence and loss aversion are phylogenetically ancient enough to be shared with capuchin monkeys—as our work suggests—it's unlikely that the human species will overcome these tendencies anytime soon.

How, then, should we deal with the fact that our choices are at the mercy of deeply ingrained irrational strategies? One way, advocated by the behavioral economist Richard Thaler, is to harness these biases for our benefit.[6] We may

---

6 R. H. Thaler and C. R. Sunstein, *Nudge: Improving Decisions on Health, Wealth, and Happiness* (New Haven, Conn.: Yale University Press, 2008).

be at the mercy of reference dependence, but there's lots of flexibility in what counts as a reference point. Using subtle changes in wording and framing, we can switch how we instinctively think about a problem and make the most rational option feel more intuitive. Thaler has used this idea to develop a better retirement savings plan, one that increases people's savings contributions automatically after they've received a pay raise. By taking retirement contributions before people have a chance to adjust to their new paycheck's reference point, Thaler's plan avoids loss aversion and allows people to feel better about saving more.[7] Similar reference-point changes have been used to increase other good behaviors. The psychologist Noah Goldstein observed that hotel guests are more likely to reuse bath towels when they are informed that most previous guests chose to do so. The actions of others provide a powerful reference point against which we strive to avoid seeming less environmentally correct.[8]

With newfound insights about the phylogenetic origins of our irrational decision-making strategies in place, social scientists are now poised to discover new ways we can harness our evolved biases to further modern decision-making agendas—such as making better financial choices and per-

---

7 R. H. Thaler and S. Bernartzi, "Save More Tomorrow: Using Behavioral Economics to Increase Employee Saving," *Journal of Political Economy* 112 (2004), S164–87. For other helpful suggestions about how to use your biases to your advantage, see Thaler's blog, http://nudges.wordpress.com/.

8 N. J. Goldstein, R. B. Cialdini, and V. Griskevicius, "A Room with a Viewpoint: Using Social Norms to Motivate Environmental Conservation in Hotels," *Journal of Consumer Research* 35 (2008), 472–82.

haps even increasing our happiness. Even professional golfers have made some headway in this regard. Reference dependence may have cost Tiger a win at the 2009 Barclays, but it also gave him a way to feel better about his poor performance. When interviewed about his score on the eighteenth, Tiger was quick to highlight an alternative reference point for the press: his final putt wasn't the worst final putt in the tournament. "It's frustrating when you misread a putt that bad," he said. But one of the players he tied with "did the same thing. His putt broke more." Changing your reference point may be an evolutionarily old strategy, but it's also a smart one. And as any golfer can tell you, if you look carefully you can always find a worse putt.

SAMUEL M. McCLURE
received his PhD in neuroscience from Baylor College of Medicine in 2003. After a postdoctoral stint at Princeton University, he joined the psychology faculty at Stanford University as an assistant professor in 2007. McClure's work has combined behavioral, computational, and neuroimaging methods to investigate the neural basis of reward processing and decision making. More recently, he has focused on the neural mechanisms of delay discounting—the processes by which we evaluate goods available in the future.

OUR BRAINS KNOW WHY WE DO WHAT WE DO

SAMUEL M. McCLURE

Decision making is broadly understood to result from the interaction of cognitive heuristics (fast, automatic responses) with rule-based, deliberative processes. How we choose at any given time depends critically on our automatic evaluations and on which of our numerous decision heuristics we draw upon to formulate a judgment. But our decisions are also determined in part by how ready and willing we are to pause and use reasoning to systematically weigh the costs and benefits of each alternative.

Thanks to recent advances in neuroscience, our understanding of brain function will increasingly refine this conceptualization of the decision-making process. The wide arsenal of heuristics that have been identified will be reduced significantly, since many draw on the same neural substrates. Additionally, the nature of the interaction between deliberate and automatic processes will be illuminated by identifying the brain systems that support those processes. The goal of the new field of decision neuroscience is a greatly improved understanding of the variability that dominates our moment-to-moment decision-making behavior.

A lot rides on that understanding. How, for example, can you induce people to give to worthy charities? Is it possible to educate people to be less susceptible to scam artists (conservatively a $6 billion annual industry)? Economists and policy makers need reliable methods for estimating choices in order to design effective policies to promote retirement savings and simultaneously encourage enough spending to maintain economic growth. The actions of the Federal Reserve provide a good illustration of the importance of decision science. Determining whether and how much to change short-term interest rates depends on estimates of how a contemplated change will affect decisions to borrow and, following a complex cascade of resulting decisions by many individuals and corporations, how this will affect our nation's economy. Without comprehensive models of how people incorporate incentives when making decisions, it would be nearly impossible to create well-founded public policies, or at least efficient ones.

For the most part, models of decision making that inform public policy are dominated by the notion that people behave as perfectly rational agents. If we assume that people always act to maximize the good that befalls them, establishing policies is greatly simplified. If you want to dissuade people from smoking cigarettes, increase their cost and people will buy fewer packs. Retirement savings will be built up by IRA and 401(k) plans that provide tax benefits for money saved. Even the concern of public health officials about the expanding waistlines of most Americans has led to proposals for taxes on sugar-containing drinks.

Despite the simplicity and elegance of the rational-agent model, however, it is woefully inadequate when it comes to describing many of the intricacies of how we make decisions. One example is known as the energy-efficiency gap. For some time now, technology has been available to make refrigerators, air conditioners, and other household appliances much more energy-efficient. The problem is that these technologies come at a cost; refrigerators that are more energy-efficient are also more expensive. This cost differential can generally be made up by reduced electric bills accumulated over some period of time. Yet people tend to buy energy-efficient appliances only if the time between the purchase and the recouping of the up-front cost is extremely short. If this period is converted to a number equivalent to a savings account interest rate, most people require an annual rate somewhere between 40 and 100 percent before they will consider buying an energy-efficient refrigerator. This rate is, of course, well outside what any normative model would consider rational, giving rise to a gap in energy-efficient purchases that has been hard to close.

The recognition that people deviate widely from rational models is nothing new; it has provided fodder for fiction for centuries. However, the development of quantitative models superior to rational-agent theory is still relatively recent. In the 1950s, the protean social scientist Herbert

Simon devised the concept of bounded rationality to capture one fundamental attribute of our decision-making
behavior. We have imperfect memories and limited time in
which to consider and weigh all the attributes of the choices
available to us. It may be presumed that better refrigerators would be bought if people were to use financial calcu-
lators when evaluating different models. However, taking
the time to make choices in this way (and learning how to
convert price differences to interest rates in the first place)
is too burdensome to be practical. Generally, we tend to
make decisions using simpler strategies, at the cost of occasionally suboptimal results. Simon described one strategy
that people use when making complex choices—a strategy
he termed "satisficing." Satisficing is the "good enough"
strategy. We may begin by wanting a refrigerator that is
large enough to store all the food for the family but is also
Energy Star rated. Once we find a model that fits these two
criteria, we buy it, even though it may not have all the features we would like.

Satisficing is just one example of many fast-and-frugal
strategies we tend to employ when making choices. We use
other shortcuts in decision making as well. The psychologists Amos Tversky and Daniel Kahneman, beginning in
the 1970s, uncovered numerous decision heuristics we use
in making judgments. For example, when estimating probability, we sometimes use the availability heuristic; that is,
we base our estimate of the probable frequency of some
event on how easily we can remember past instances. This
leads to overestimation of such matters as the likelihood
of suffering a shark attack, an event commonly sensation-
alized in the media. Our judgments are also subject to

many biases. We tend to be risk-averse, meaning that we are overly pessimistic about probable outcomes. We also misuse probabilities, tending to be overly optimistic about low-probability events (e.g., we buy lottery tickets) and overly pessimistic about high-probability events (e.g., we worry too much about plane crashes). Combining much of this work, Kahneman and Tversky formulated prospect theory, a remarkably successful model of how we make choices.

Prospect theory describes some cases in which our decisions can have profound and disturbing consequences on our behavior. For example, we tend to be loss-averse, meaning that losses loom larger in our imaginations than do equivalent gains. For activities such as gambling, this is not too large a problem. When the odds are fifty-fifty, people generally find a gamble attractive if the potential benefits are at least twice as large as the potential losses. Where loss aversion becomes troubling is in its consequences for some important decisions, such as deciding between possible medical treatments. If a given procedure is explained in loss-related terms, we evaluate it very differently than when it is framed in gain-related terms. The prospect of surgery is much less attractive when it is described as involving a 1 percent chance of dying on the operating table than when it is described as having a 99 percent chance of survival. This is true despite the fact that the two descriptions are equivalent. It is true both for patients considering the procedure and doctors trying to determine whether or not to recommend it. One would hope that how we behave would not be sensitive to such trivial variations.

Although decision heuristics can sometimes lead us

astray, they are generally adaptive in that they lead to good decisions most of the time. The German psychologist Gerd Gigerenzer has demonstrated this in numerous experiments, and the notion was recently popularized in Malcolm Gladwell's book *Blink,* which illustrates the efficacy of split-second decision making. It is generally agreed that heuristics are adaptive because they are slowly learned from lots of experience over time. For example, the availability heuristic most of the time produces the correct conclusion; we rely on memory to estimate frequency because it does so accurately (most of the time).

But there are two related problems with a heuristic-based model of decision making. First, we can never know whether all heuristics have been discovered. There is no overarching theory for what leads to the development and use of a heuristic. Individual heuristics must be intuited and tested. For similar reasons, we cannot know when heuristics are redundant. Gigerenzer describes one heuristic as "When you see a white coat, trust it" to capture the fact that we trust doctors despite the fact that they are subject to their own decision fallacies. Is this truly an independent decision heuristic, or is it a specific instance of a more general heuristic? What is the more general heuristic? And may the same question be applied to it?

A second problem with current models of decision making lies in the variability of individual properties of judgment and valuation, such as risk aversion and loss aversion, which are known to depend on numerous contextual factors. Mood and incidental emotions have been a particular focus of studies investigating this variability. The degree of your loss aversion depends on whether you're feeling happy or sad. Memory recall is easier for events consistent with your current mood, which suggests that the availability heuristic may be differentially applied. This means that arbitrary circumstances such as the weather can alter how we decide at any given moment. The consequences of contextual effects can be profound. The stock market, for example, is known to perform better on sunny days than when the weather in lower New York City is inclement.[1] We are all aware of these effects in our day-to-day lives. Be wary about going to the supermarket when you're really hungry, unless you want to buy a lot of food. Don't send e-mails when you're angry; your judgments will change when you're in a more placid frame of mind. The problem with our current understanding of decision heuristics is that they don't include emotions in their formulation. To include these effects, new theories have to be developed. Again, without a more general theory of decision making, understanding the full nature of contextual dependencies depends entirely on intuition and experimentation.

A theory that encompassed the underlying mechanisms that generate choices would reveal gaps where potential undiscovered heuristics may exist. Effects of context and mood would become more apparent. The clear path to developing such a model is by describing behavior in terms of the brain.

The main challenge for developing a brain-based description of decision making has been observing brain function. Experiments in animals over the past several

---

1 D. Hirshleifer and T. Shumway, "Good Day Sunshine: Stock Returns and the Weather," *Journal of Finance* 58 (2003), 1009–32.

decades have identified numerous brain systems involved in evaluation and choice. However, animal models have only limited applicability to people. They may be used for developing theories of brain function, but human behavior and neuroanatomy are significantly different from those of every other species. Moreover, the tools used for animal studies, such as implantation of electrodes, are generally proscribed for use on human subjects. Until recently, this meant that in order to investigate human brain function, we had to rely on electroencephalography, in which electrical activity is measured at the scalp. However, EEGs are unable to record from the deep brain areas associated with valuation and reward processing. Lesion studies, in which brain function is inferred from the loss of function following brain injury, have provided insight, but these studies suffer from other limitations. The field of decision neuroscience took off only in the 1990s, with the development of functional magnetic resonance imaging, or fMRI.

Functional MRI is not a perfect technique. It measures signals related to blood flow in the brain and not the activity of neurons directly. However, it is able to measure equally well in virtually every brain region. Measuring blood flow is not as haphazard as it may sound. When neurons are firing action potentials to signal other neurons, they use energy. Blood flow increases locally in response to this energy demand. The precise relationship underlying this response is still being worked out, but the fact that blood flow increases in localized regions correlated with local neuronal activity is unquestioned. For this reason, fMRI is a powerful surrogate method for measuring brain activity in real time as people perform any task that can be performed in the middle of a large cylindrical magnet.

With numerous brain systems involved in reward processing and decision making already known, the next step was to test with fMRI how these brain systems relate to loss aversion, risk aversion, and the other parameters of valuation that constitute prospect theory. The outcome of these experiments was astounding. In study after study, all of these various parameters, which describe individual differences in valuation, correlate with brain activity in the same limited number of brain structures: the striatum, amygdala, ventromedial prefrontal cortex, and insula. Loss aversion has been found to be fully accounted for by responses in the striatum, though occasionally the insula is activated—which is potentially interesting because the insula is associated with emotions, notably disgust. This may explain why loss aversion varies from situation to situation. If feelings of disgust are implicated, the brain response changes and valuation may be expected to change as well. Probability and risk aversion have also been found to depend in part on responses in the striatum. This suggests that loss and risk aversion, two seemingly independent constructs, may be related in important ways that were unanticipated on the basis of behavioral observations alone.

The surprising finding that the various complexities of decision theories all map to a limited set of brain systems has profound potential consequences. Factors that alter decision tendencies, such as mood, may now be easily understood. Other work has shown that emotions map to the same neural systems, so if mood alters responses in these brain regions, decisions should be altered as well. This has led to some exciting new predictions that have now been tested. For example, it is known that the striatum responds to stimuli promising reward, such as sexually alluring photographs.[2] This striatal reaction undoubtedly carries over to affect our responses to risky choices.

Perhaps more important, the incorporation of neuroscience into the study of decision making may change how we define patterns of behavior. Rather than being labeled "timid" or "cautious," a person may be described as having elevated amygdala sensitivity, which implies greater risk aversion in some circumstances and greater loss aversion in others and so forth. The transition to neuroscientific terminology will be necessary once it is understood how various behaviors interrelate through brain function.

One way in which people differ markedly from other animals is in our ability to respond flexibly and rationally. We can think abstractly about a situation and call upon symbolic systems, such as language, to represent the properties of a choice so as to help us make decisions. Neuroscience will contribute in critical ways to understanding these cognitive processes as well. As with the more basic brain functions, such as reward processing, work has identified the brain systems involved in cognitive deliberation. In general, these processes rely on the lateral prefrontal cortex and posterior parietal cortex. Disruptions of activity in these regions impair our ability to adapt our behavior

---

2 B. Knutson et al., "Nucleus Accumbens Activation Mediates the Influence of Reward Cues on Financial Risk Taking," *Neuroreport* 19 (2008), 509–13.

to the specific demands of a situation. This loss of function lies at the heart of many psychiatric disorders, such as schizophrenia, and has therefore long been the subject of neuroscientific study.

For our purposes here, it is important to note that the lateral prefrontal cortex connects to each of the above-mentioned brain structures implicated in reward processing. This suggests the existence of a mechanism by which we can deliberately control how we evaluate choices—and thereby control our decision-making behavior cognitively. One exciting study of cognitive involvement in decision making was recently conducted by the Caltech neuroeconomists Todd Hare, Colin Camerer, and Antonio Rangel. They were interested in how dieters and control subjects balanced nutrition and taste when deciding what to eat. They found (as expected) that abstract information, such as nutritional values, is encoded by the lateral prefrontal cortex, whereas taste information is represented in the striatum and ventromedial prefrontal cortex. For the difficult choices involving food that is not particularly appealing but is nutritionally valuable (think apple as opposed to potato chips), whether nutrition wins depends on whether the lateral prefrontal cortex is able to alter activity in the ventromedial prefrontal cortex. The study's participants had to willfully change their valuations, and this process could be directly observed in their ongoing brain activity.[3]

Deliberation and willpower are fundamental properties

3 T. A. Hare, C. F. Camerer, and A. Rangel, "Self-Control in Decision-Making Involves Modulation of the vmPFC Valuation System," *Science* 324 (2009), 646–48.

of decision making. Our willingness to use such processes has a profound effect on our choices. We have already seen this in the case of the energy-efficiency gap. However, the nature of this interaction is still mostly mysterious. There is some evidence that willpower functions like a muscle; that is, if you use it a lot doing one thing—say, completing a demanding task—fatigue sets in and less willpower will be exerted for some period of time afterward. In our daily behavior, we navigate the world primarily by using automatic mechanisms—at least until something seems wrong, at which point we may or may not stop and think more carefully. By measuring each of these processes, neuroscience stands to revolutionize the way we understand self-control, with profound consequences for our understanding of decision making.

Decision neuroscience is an emerging discipline that lies at the intersection of two rapidly advancing fields. With prospect theory at its helm, it provides quantitative models of human behavior far superior to any others that have ever existed. On the physiological side, neuroscience now allows for direct recording from the human brain in ways never before possible. The fusion of these two sciences has produced exciting results that only hint at what is to come. Decision making will eventually be understood in neuroscientific terms, leading to much more generalizable theories and a far better understanding of such elusive constructs as willpower and self-control.

JENNIFER JACQUET
graduated with a master's degree in environmental economics from Cornell University in 2004 and earned a PhD in 2009 from the University of British Columbia, where she now holds a postdoctoral fellowship. As part of the Sea Around Us Project, a joint collaboration between the university and the Pew Charitable Trusts, she researches market-based conservation initiatives related to seafood and other natural resources. With colleagues from the Max Planck Institute for Evolutionary Biology and UBC's Mathematics Department, she is currently conducting a series of games and experiments to study the effects of honor and shame on cooperation.

IS SHAME NECESSARY?

JENNIFER JACQUET

Financial executives received almost $20 billion in bonuses in 2008 amid a serious financial crisis and a $245 billion government bailout. In 2008, more than 3 million American homes went into foreclosure because of mortgage blunders those same executives helped facilitate. Citigroup proposed to buy a $50 million corporate jet in early 2009, shortly after receiving $45 billion in taxpayer funds. Days later, President Barack Obama took note in an Oval Office interview. About the jet, he said, "They should know better." And the bonuses, he said, were "shameful."

What is shame's purpose? Is shame still necessary? These are questions I'm asking myself. After all, it's not just bankers we have to worry about. Most social dilemmas exhibit a similar tension between individual and group interests. Energy, food, and water shortages, climate disruption, declining fisheries, increasing resistance to antibiotics, the threat of nuclear warfare—all can be characterized as tragedies of the commons, in which the choices of individuals conflict with the greater good.

Balancing group and self-interest has never been easy, yet human societies display a high level of cooperation. To attain that level, specialized traits had to evolve, including such emotions as shame.[1] Shame is what is supposed to occur after an individual fails to cooperate with the group. Shame regulates social behavior and serves as a forewarning of punishment: conform or suffer the consequences. The earliest feelings of shame were likely over issues of waste management, greediness, and incompetence. Whereas guilt is evoked by an individual's standards, shame is the result of group standards. Therefore, shame, unlike guilt, is felt only in the context of other people.

The first hominids could keep track of cooperation and defection only by firsthand observation. Many animals use visual observations to decide whether to work with others. Reef fish in the Red Sea, for instance, watch wrasses clean other reef fish to determine whether or not they're cooperative, as the biologist Redouan Bshary discovered. Bshary went scuba diving off Egypt's coast to observe this symbiotic relationship. Bluestreak Cleaner wrasses (*Labroides dimidiatus*) eat parasites, along with dead or infected tissue, off reef fish in more than two thousand interactions

1 R. Boyd and P. J. Richerson, "Culture and the Evolution of Human Cooperation," *Proceedings of the Royal Society of London B* 364 (2009), 3281–88.

a day, each of which can be considered an act of cooperation. Wrasses are tempted to eat more than just the parasites, but if the reef fish loses too much flesh in the deal, it
will refuse to continue working with the wrasse. Reef fish
approach wrasses that they see cooperating with their current clients and avoid the wrasses they see biting off more
than they should.[2]

Like the Bluestreak Cleaner wrasses, humans are
more cooperative when they sense they're being watched.
Researchers at the University of Newcastle upon Tyne
examined the effect of a pair of eyes on payments for tea
and coffee to an honesty box. Alternating images of flowers and human faces were posted above the box in the university coffee room each week for ten weeks; researchers
found that people paid nearly three times as much for their
drinks in weeks during which they were exposed to the
human gaze.[3]

The feeling of being watched enhances cooperation, and
so does the ability to watch others. To try to know what
others are doing is a fundamental part of being human. So
is fitting in. The more collectivist the human society, the
more important it is to conform and the more prominent
the role of shame.[4] Shame serves as a warning to adhere

2 R. Bshary, "Biting Cleaner Fish Use Altruism to Deceive Image-
  Scoring Client Reef Fish," *Proceedings of the Royal Society of
  London B* 269 (2002), 2087–93.

3 M. Bateson, D. Nettle, and G. Roberts, "Cues of Being Watched
  Enhance Cooperation in a Real-World Setting," *Biology Letters*
  2 (2006), 412–14.

4 D. M. T. Fessler, "Shame in Two Cultures: Implications for Evolutionary Approaches," *Journal of Cognition and Culture* 4 (2004), 2.

to group standards or be prepared for peer punishment. Many individualistic societies, however, have migrated away from peer punishment toward a third-party penal system, such as a hired police force, formal contracts, or trial by jury. Shame has become less relevant in societies where taking the law into one's own hands is viewed as a breach of civility.

Perhaps this is why it makes us uncomfortable to contemplate shaming people: shame invites the public in on the punishment. Consider the stocks, scab lists during union strikes, or Nathaniel Hawthorne's *The Scarlet Letter*. Or the proposal made by the prominent conservative William F. Buckley, Jr., in 1986 to tattoo people with AIDS. These instances of shaming now seem an affront to individual liberty. Getting rid of shaming seems like a pretty good thing, especially in regulating individual behavior that does no harm to others. In eschewing public shaming, society has begun to rely more heavily on individual feelings of guilt to enhance cooperation.

Guilt prevails in many social dilemmas, including one area of my own research: overfishing. At the root of the problem of overfishing is the human appetite. Wild-fish catches are declining, and many of us seek to avoid the guilt brought on by eating unsustainable seafood. Here are just a few recent headlines from major newspapers: HOLY MACKEREL AND OTHER GUILT-FREE FISH (*The New York Times*), GUILT-FREE SUSHI (*The Christian Science Monitor*), COD AND CHIPS? MAKE IT POLLOCK IN GUILT-FREE GUIDE TO SEAFOOD (*The Times of London*), and A GOOD APPETITE; SEAFOOD, EASY AND GUILT-FREE (*The New York Times*).

It is perhaps unsurprising that a set of tools has emerged to assuage this guilt and, in the case of seafood, reform the appetite.[5] These tools aim to divert demand from one type of seafood toward another. Wallet cards, iPhone apps, and ecolabels tell consumers which fish ought to be and ought not to be eaten. Shoppers in Europe have been given rulers so that they can measure fish and avoid buying juveniles.

Guilt abounds in many situations where conservation is an issue. Harried by guilt, one mother reuses her daughter's bathwater for her own bath. Los Angeles shoppers refuse to buy blueberries imported from Chile because of the fuel consumed in shipping them. Another woman feels guilty about the natural habitat lost to cocoa cultivation and refuses to buy chocolate, prompting her husband to say that she took the joy out of his Almond Joy.[6] Just as the devout purchased guilt-alleviating papal indulgences in the Middle Ages, guilt-ridden consumers today buy carbon offsets, LED lightbulbs, and hybrid cars and can be guided to something approaching sanctity by books such as *The Virtuous Consumer, The Rough Guide to Shopping with a Conscience,* and *The Eco Chick Guide to Life: How to Be Fabulously Green.*

The problem is that environmental guilt, though it may well lead to conspicuous ecoproducts, does not seem to elicit conspicuous results. One supermarket chain introduced signs at the fish counter to show the most and

---

5 J. Jacquet et al., "Conserving Wild Fish in a Sea of Market Based Efforts," *Oryx* 44 (2010), 45–56.

6 C. Crawford, "Green with Worry," *San Francisco Magazine*, February 2008.

least sustainable seafood: Sales of the green-tagged "best choice" fish increased an average of 29 percent per week, sales of yellow-tagged "proceed with caution" seafood declined an average of 27 percent per week, but the sales of the red-tagged "worst choice" seafood—i.e., the heavily overfished species—remained the same.[7] Between 1980 and 2008, sales of pesticides increased by 36 percent in the state of California, the birthing ground of the organic food ecolabel.[8] Despite sporadic instances of such measures as carpooling and the use of cloth grocery bags in lieu of plastic, the demand for oil in the United States has grown by 30 percent overall and 5 percent per capita since 1990.[9] The positive effect of idealistic consumers does exist, but it is masked by the rising demand and numbers of other consumers.

Guilt is a valuable emotion, but it is felt by individuals and therefore motivates only individuals. Another drawback is that guilt is triggered by an existing value within an individual. If the value does not exist, there is no guilt and hence no action (e.g., the sales of red-tagged "worst choice" seafood remained the same). What if the aim were to promote a value felt by the group but not necessarily by every individual in the group? Many problems, like most concerning the environment, are group problems. Perhaps

---

7 E. Hallstein and S. B. Villas-Boas, "Are Consumers Color Blind? An Empirical Investigation of a Traffic Light Advisory for Sustainable Seafood," www.escholarship.org/uc/item/29v6w5sp#page-11.

8 www.cdpr.ca.gov/docs/mill/sumpdsld.pdf.

9 This is calculated from Department of Energy statistics. U.S. oil consumption in 1990 was 17 million barrels per day and in 2010 was 22.2 million barrels per day.

to solve these problems we need a group emotion. Maybe we need shame.

Shaming, as noted, is unwelcome in regulating personal conduct that doesn't harm others. But what about shaming conduct that does harm others? The U.S. National Sex Offender Registry provides an online database with the names, photographs, and addresses of sex offenders in every state. In March 2010, Nebraska lawmakers approved a bill that allows the state to publish online the names and addresses of people owing more than $20,000 in taxes. Judges in various states issue shaming punishments, such as sentencing pickpockets and robbers to carry picket signs that announce their crimes to the public. These instances of shaming might deter bad behavior, but critics such as Martha Nussbaum, a political philosopher at the University of Chicago, argue that shaming by the state conflicts with the law's obligation to protect citizens from insults to their dignity.[10]

What if government is not involved in the shaming? A neighborhood in Leicester, England, has a YouTube channel dedicated to neighborhood issues, including catching "litter louts." A collection of videos shows individuals caught in various acts of littering, and if someone recognizes the litter lout, he or she can e-mail the lout's identity to the neighborhood management board, which passes it on to the City Council so that fines can be issued and the video removed.[11] In 2008, *The Santa Fe Reporter*

10 M. Nussbaum, *Hiding from Humanity: Disgust, Shame, and the Law* (Princeton, N.J.: Princeton University Press, 2004).

11 http://stpetersnm.com/litter_louts.html.

published the names and addresses of the top ten water-using households in the city (first place went to a home owner who used twenty-one times the household average). The tennis club near my apartment in Vancouver, British Columbia, publishes the names of people who don't pay their dues. In each of these cases, the activity of the individual compromises the community. In none of them is the state involved in the shaming. Is this a fair use of shaming? Is it effective?

Let's deal with the latter question. Shaming might work to change behavior in these cases, but in a world of urgent, large-scale problems, changing individual behavior is insignificant. Small changes, adopted by one individual at a time, can make a difference in a problem only when the problem is small or there is lots of time to solve it (for instance, in marginalizing politically incorrect words). Many of today's social movements, like the industries they seek to revolutionize, must make big changes quickly—which is best accomplished by directing efforts upward toward institutions. I call this vertical agitation. *The Santa Fe Reporter* listed the top ten commercial water users, in addition to the top ten households. The first of these offenders, the city of Santa Fe, used 195 times as much water as the number one household offender. Imagine the relative difference in getting the city to commit to water-saving techniques as compared to reforming a single household.

Guilt cannot work at the institutional level, since it is evoked by individual scruples, which vary widely. But shame is not evoked by scruples alone; since it's a public sentiment, it also affects reputation, which is important to an institution. At the 2004 meeting of the World Economic Forum in Davos, Switzerland, leading CEOs issued a press release showing that corporate brand reputation outranked financial performance as the most important measure of success. For an example of how shame and reputation interact, consider restaurant hygiene cards, introduced in 1998 by Los Angeles County as a shaming technique in the interests of public health. Restaurants were required to display grade cards that corresponded to their most recent government hygiene inspection. The large grade in the window—A, B, or C—honors restaurants that value cleanliness most and shames those that value it least. The grade cards have apparently led to increased customer sensitivity to restaurant hygiene, a 20 percent decrease in countywide hospitalizations for food-borne illnesses, and better hygiene scores for county restaurants.[12]

Recall that in our early evolution we could gauge cooperation only firsthand. As group size got bigger and ancient humans grappled with issues of necessary cooperation, the human brain became better able to keep track of all the rules and all the people. The need to accommodate the increasing number of social connections and monitor one another could be, according to the social grooming hypothesis put forward by the British anthropologist Robin Dunbar, why we learned to speak.[13] Then, five thousand years ago, there arose another tool: writing. Lan-

---

12 G. Zhe and P. Leslie, "The Case in Support of Restaurant Hygiene Grade Cards," *Choices* 20 (2005), 97–102.

13 See especially his *Grooming, Gossip, and the Evolution of Language* (Cambridge, Mass.: Harvard University Press, 1997).

guage, both oral and written, allowed for gossip, a vector of social information. Research carried out by Ralf Sommerfeld of the Max Planck Institute for Evolutionary Biology and his colleagues demonstrated that in cooperation games that allowed players to gossip about one another's performance, positive gossip resulted in higher cooperation. Of even greater interest, gossip affected the players' perceptions of others even when they had access to first-hand information.[14]

Human society today is so big that its dimensions have outgrown our brains. We have an increasing number of people and norms. What tool could help us gossip in a group this size? Nowadays we keep track of and distribute unprecedented amounts of information via our computers: for example, journalists, interest groups, and ordinary citizens can access the U.S. Environmental Protection Agency's Toxics Release Inventory database online to identify and shame polluters. Between the database's inception in 1988 and 1995, releases of 330 toxic chemicals on the list have declined by 45 percent.[15] After the retailer Trader Joe's was unresponsive to requests by the nonprofit group Greenpeace to stop selling unsustainable seafood, Greenpeace coordinated singing-fish telephone calls or demonstrations at every Trader Joe's across the nation, using the

---

14 R. H. Sommerfeld et al., "Gossip as an Alternative for Direct Observation in Games of Indirect Reciprocity," *Proceedings of the National Academy of Sciences* 104 (2007), 17435–40.

15 A. Fung and D. O'Rourke, "Reinventing Environmental Regulation from the Grassroots Up: Explaining and Expanding the Success of the Toxics Release Inventory," *Environmental Management* 25 (2000), 115–27.

Internet. The CEO of Trader Joe's decided to comply with Greenpeace's demands by dropping several overfished species and agreeing to sell only sustainable seafood by the end of 2012.

We can use computers to simulate some of the intimacy of tribal life, but we need humans to evoke the shame that leads to cooperation. The emergence of new tools—language, writing, the Internet—cannot completely replace the eyes. Face-to-face interactions, such as those outside Trader Joe's stores, are still the most impressive form of dissent.

So what is stopping shame from catalyzing social change? I see three main drawbacks:

1. *Today's world is rife with ephemeral, or "one-off," interactions.* When you know you're unlikely to run into the same situation again, there is less incentive to change your behavior. Research shows, however, that if people know they will interact again, cooperation improves.[16] Shame works better if the potential for future interaction is high. In a world of one-off interactions, we can try to compensate for anonymity with an image score, such as hygiene grade cards or eBay's seller ratings, which sends a signal to the group about an individual's or institution's degree of cooperation.

2. *Today's world allows for amorphous identities.* Recall the reef fish that observe Bluestreak Cleaner wrasses in the Red Sea. The wrasses seem to know they are being

---

16 M. Milinski, D. Semmann, and H. Krambeck, "Reputation Helps Solve the 'Tragedy of the Commons,'" *Nature* 415 (2002), 424–26.

watched, and certain wrasses build their reputation on the small reef fish, allowing the big reef fish to observe their cooperative behavior with the small fry. Then, when the big fish comes in for its own cleaning, these wrasses eat some of the big reef fish's flesh along with its parasites, fattening themselves on their defection. To add to the confusion on the reef, False Cleanerfish (*Aspidontus taeniatus*) make their living by looking very similar to the Bluestreak Cleaner wrasses. They are able to approach reef fish under the guise of cooperation and then bite off pieces of fish flesh and swim away.

Many of our interactions these days are similar to the fish cleanings in the Red Sea. It's hard to keep track of who cooperates and who doesn't, especially if it's institutions you're monitoring. Enron, which in 2001 filed one of the largest bankruptcies in U.S. history, hid billions of dollars in debt in hundreds of shell firms, which bought poorly performing Enron stocks so that Enron could create a fraudulent company profile and mislead its auditors. Lehman Brothers, in the years before its 2008 collapse, used a smaller firm called Hudson Castle (of which it owned 25 percent) to shift risky investments off its books so that Hudson Castle, not Lehman Brothers, could absorb the "headline risk." Which leads us to shaming's third weakness.

3. *Shaming's biggest drawback is its insufficiency.* Some people have no shame. In the research my colleagues and I have conducted on first-year students involving games that require cooperation, we have found that shame does not always encourage cooperation from players who are least cooperative. This suggests that a certain fraction of

a given population will always behave shamelessly, like the False Cleanerfish, if the payoff is high enough. The banks may have gone bankrupt, but the bankers got their bonuses. There was even speculation that publishing individual bankers' bonuses would lead to banker jealousy, not shame.

My colleagues and I conclude that ultimately shame is not enough to catalyze major social change. Slavery did not end because abolitionists shamed slave owners into freeing their slaves. Child labor did not stop because factories were shamed into forbidding children to work. Destruction of the ozone layer did not slow because industries were ashamed to manufacture products that contained chlorofluorocarbons. This is why punishment remains imperative. Even if shaming was enough to bring the behavior of most people into line, governments need a system of punishment to protect the group from the least cooperative players.

Finally, consider who belongs to the group. Today we are faced with the additional challenge of balancing human interests and the interests of nonhuman life. How can we encourage cooperation among all living things when the nonhumans have no voice? Successful species will likely be those that recognize, implicitly or explicitly, life's interdependency. If humans are to succeed as a species, our collective shame over destroying other life-forms should grow in proportion to our understanding of their various ecological roles. Maybe the same attention to one another that promoted our own evolutionary success will keep us from failing the other species in life's fabric and, in the end, ourselves.

KIRSTEN BOMBLIES,
born in Germany and brought up in Colorado, is a molecular biologist and an assistant professor in the Organismic and Evolutionary Biology Department at Harvard University. She has a BA in biochemistry and biology from the University of Pennsylvania and a PhD in genetics from the University of Wisconsin (2004), where she worked with John Doebley. From 2004 to 2009, she was a postdoc in Detlef Weigel's group at the Max Planck Institute for Developmental Biology in Tübingen, Germany, where she "uncovered my current passions" for plant speciation, immune-system evolution, adaptation, and biogeography. In 2008, she received a MacArthur Fellowship.

PLANT IMMUNITY IN A CHANGING WORLD

KIRSTEN BOMBLIES

Our survival on Earth depends crucially on plants. Almost all our food comes directly or indirectly from several dozen major crop species cultivated around the globe. Many of our building materials are derived from plants—not to mention the oxygen in the air we breathe. But many of the forest, grasslands, and crop species that we and most of the planet's other organisms depend on are increasingly in peril because of habitat degradation, pollution, and climate change. These factors also threaten to dramatically alter the distribution and prevalence of pathogens, herbivores, and competitors. How crop and wild-plant species will fare in the face of such challenges combined with other stresses associated with rapid climate change is of great concern; not least among the worries is how disease patterns may change. Models of the effects of climate change predict not only a rise in global average temperature but also greater frequency and severity of extreme weather events, which have already been implicated as possible major causes of increased pest damage and crop losses in recent decades.[1] To what degree plant populations can adapt to novel disease pressures in an altered and increasingly unpredictable climate remains largely unknown.

Though potential threats to plants are numerous, there is reason for hope. Some plant species are hardy—resilient to shifts in climate and tolerant of short-term stress. Such species may well survive climate or habitat changes largely unscathed, providing us with valuable information about resistance to climate-related problems and serving as repositories of genetic solutions. For plants that are less tolerant, we may yet be able to help with informed interventions or conservation strategies. In some cases, change may have mixed effects. For example, elevated temperature may cause heat stress but reduce pathogen virulence in summer while alleviating cold stress but promoting pathogen survival in winter.[2] Information is critical. Before we can anticipate effects or consider interventions, we need to improve our plant's-eye view of the world.

A major factor dominating plant lives is, of course, that they are sessile—rooted in place. Thus, they must deal

---

1 See, for example, C. Rosenzweig et al., "Climate Change and Extreme Weather Events," *Global Change and Human Health* 2 (2001), 90–104; and K. A. Garrett et al., "Climate Change Effects on Plant Disease: Genomes to Ecosystems," *Annual Review of Phytopathology* 44 (2006), 489–509.

2 Garrett et al., "Climate Change."

with their local environment and its vagaries; escape is not an option. They have thus evolved effective ways of sensing their surroundings and responding appropriately. They can adjust their development and reproductive timing and shift resource allocation in response to cues as diverse as crowding; attack by pathogens or herbivores; the availability of nutrients, light, and moisture; and changes in temperature.[3] Most plant species are well adapted to the range of conditions that commonly occur in their natural habitats; when conditions stray far from optimal, however, such as during drought, heat waves, or floods, stress responses that help the plant survive and minimize damage are activated. For example, drought stress can induce plants to close gas-exchange pores in their leaves to limit water loss, water submergence can induce stem elongation in flood-tolerant species to help them reach air, and pathogen attack can induce hardening of cell walls and production of protective compounds.[4]

A confounding problem arises when significant stresses are combined; indeed, it is the combining of stresses that is particularly lethal to crop plants (a problem well

3 M. A. Jenks and P. M. Hasegawa, eds., *Plant Abiotic Stress* (Ames, Iowa: Blackwell, 2005).

4 See, for example, C. García-Mata and L. Lamattina, "Nitric Oxide Induces Stomatal Closure and Enhances the Adaptive Plant Responses Against Drought Stress," *Plant Physiology* 126 (2001), 1196–204; L. A. C. J. Voesenek et al., "Interactions Between Plant Hormones Regulate Submergence-Induced Shoot Elongation in the Flooding-Tolerant Dicot *Rumex palustris*," *Annals of Botany* 91 (2003), 205–11; and R. Hückelhoven, "Cell Wall–Associated Mechanisms of Disease Resistance and Susceptibility," *Annual Review of Phytopathology* 45 (2007), 101–27.

known to farmers but until recently comparatively neglected by molecular biologists).[5] This is partly because some stress-response pathways antagonize each other, preventing the plant from responding effectively to more than one stress at a time. For example, plants suffering from environmental stress are often poor at fending off pathogens.[6] It is important that we have a good understanding of how and why stress-response and immune-signaling pathways interact, from the whole organism to the molecular scale, in order to predict how plants will respond when faced with combined challenges. The topic of how stress responses interact and sometimes antagonize one another is receiving extensive attention from plant biologists.[7]

What can molecular geneticists contribute to the picture? Understanding the molecular events involved will clarify why there are trade-offs in stress responses. Molecular information can help us predict how short-term responses, acclimation of individuals, or long-term adaptation of populations will be affected by environment. Knowledge of the molecular basis of responses will also enable us to assess a particular species' potential adaptability and ask whether there are genetic or physiological

5 R. Mittler, "Abiotic Stress, the Field Environment and Stress Combination," *Trends in Plant Science* 11 (2006), 15–19.

6 See, for example, L. Xiong and Y. Yang, "Disease Resistance and Abiotic Stress Tolerance in Rice Are Inversely Modulated by an Abscisic Acid–Inducible Mitogen-Activated Protein Kinase," *Plant Cell* 15 (2003), 745–59.

7 K. Yoshioka and K. Shinozaki, eds., *Signal Crosstalk in Plant Stress Responses* (Ames, Iowa: Blackwell, 2009).

constraints that limit adaptive potential or present barriers to effective intervention.

The plant immune system is complex and finely balanced, which may explain why it is sensitive to perturbation by environmental stresses. Plants deploy arsenals of physical barriers, including spines, waxes, and cell-wall modifications, as well as molecular deterrents, such as cocktails of toxic chemicals and sets of proteins, many of them similar to animal immune receptors, which can detect and signal the presence of invaders. These proteins activate a signal within plant cells that often culminates in programmed cell suicide in attacked and immediately adjacent cells. While locally damaging to the plant, this can help stall pathogens that feed on living cells (biotrophs), though it can also make the plant susceptible to pathogens that feed off of dead cells (necrotrophs).[8] An important group of surveillance proteins involved in the initial and highly specific detection of pathogens is the so-called resistance, or R, proteins. Genes encoding R proteins are common in plant genomes; often hundreds of R gene sequences are present in the genetic tool kits of individual plants.[9] From individual to individual, the sequences of these genes can be quite different. The high sequence diversity probably results from coevolution with rapidly diversifying pathogens: that is, as pathogens evolve to escape detection, new R gene specificities that can detect the novel pathogen

8 J. D. G. Jones and J. L. Dangl, "The Plant Immune System," *Nature* 444 (2006), 323–9.

9 B. C. Meyers et al., "Genome-wide Analysis of NBS-LRR-encoding genes in *Arabidopsis*," *Plant Cell* 15 (2003), 809–34.

strains may emerge and rise in frequency until the pathogen evolves a counterstrategy. This can set into motion a perpetual cycle, analogous to an arms race, that increases gene-sequence diversity in both host and pathogen. It can also confer an advantage to rare variants, because the targeted plant or the invading pathogen is less likely to have yet evolved a counterstrategy. Such mechanisms can lead to the maintenance of high levels of diversity in immune-related genes.[10]

Despite their own variability, *R* proteins activate responses through a set of signaling proteins that vary little from species to species.[11] This feature has made *R* genes very useful in plant breeding and agriculture, because it makes them interchangeable; functional resistance specificities to particular pathogens that have evolved in one context can often be successfully transferred between strains or species and still retain their resistance specificity and signaling function. This has provided excellent opportunities for plant breeders to develop crop strains resistant to common or devastating pathogens.

A general problem, which raises concerns in connection with climate change, is that pathogen signaling through *R* genes is heavily influenced by environmental cues such as temperature, humidity, salinity, soil moisture, nutrient availability, and light quality. The specific effect depends

10 J. Bergelson et al., "Evolutionary Dynamics of Plant *R*-genes," *Science* 292 (2001), 2281–85.

11 K. S. Caldwell and R. W. Michelmore, "*Arabidopsis thaliana* Genes Encoding Defense Signaling and Recognition Proteins Exhibit Contrasting Evolutionary Dynamics," *Genetics* 181 (2009), 671–84.

on the particular combination of host, stress, and pathogen. Often disease resistance and stress arising from the physical environment are antagonistic. Soybean varieties, for example, can become more sensitive to root pathogens if they are first exposed to flooding, whereas drought increases the susceptibility of peanuts to infection by a mold that produces aflatoxin, which is toxic to humans.[12] On the other hand, interactions among stress responses can also sometimes be synergistic. Exposure to salt can increase the resistance of tomato plants to the fungus *Oidium neolycopersici,* infection by various species of virus can improve drought resistance in beets, and infection of roots by the bacterium *Piriformospora indica* can boost salt tolerance in barley.[13]

These examples underline the intricate cross talk and the sometimes unexpected outcomes of interactions between environmental and pathogen stress signaling. Essential to mediating these connections is a set of signaling hormones produced by plants in response to various environmental

---

12 M. T. Kirkpatrick, J. C. Rupe, and C. S. Rothrock, "Soybean Response to Flooded Soil Conditions and the Association with Soilborne Plant Pathogenic Genera," *Plant Disease* 90 (2006), 592–6; H. R. Wotton and R. N. Strange, "Increased Susceptibility and Reduced Phytoalexin Accumulation in Drought-Stressed Peanut Kernels Challenged with *Aspergillus flavus,*" *Applied and Environmental Microbiology* 53 (1987), 270–3.

13 E. A. Achuo, E. Prinsen, and M. Höfte, "Influence of Drought, Salt Stress, and Abscisic Acid on the Resistance of Tomato to *Botrytis cinerea* and *Oidium neolycopersici,*" *Plant Pathology* 55 (2006), 178–86; P. Xu et al., "Virus Infection Improves Drought Tolerance," *New Phytology* 180 (2008), 911–21; H. Baltruschat et al., "Salt Tolerance of Barley Induced by the Root Endophyte *Piriformospora indica* Is Associated with a Strong Increase in Antioxidants," *New Phytology* 180 (2008), 501–10.

stresses. Several major players relevant to both environmental stress and pathogen signaling are salicylic acid (SA), abscisic acid (ABA), and jasmonic acid (JA). The balance of these hormones can cause either synergistic or antagonistic interactions. For example, an increase in JA in response to wounding by herbivores (especially insect pests) can inhibit SA-mediated pathogen response signaling, and ABA signaling in response to environmental stress can act either synergistically or antagonistically with SA signaling, depending on the relative concentrations of the two hormones.[14] In addition to these hormones, there are regulatory proteins that can toggle between environmental-stress tolerance and pathogen resistance pathways, activating one at the expense of the other.[15] Understanding at the molecular level how these responses interact goes a long way toward explaining some of the interaction effects that have been observed with combined stresses in field experiments.

What about temperature? Many studies have noted the temperature sensitivity of plant defense signaling mediated by *R* proteins; responses may be completely inhibited, even by otherwise nonstressful elevations in temperature.[16] Not

---

14 M. Fujita et al., "Crosstalk Between Abiotic and Biotic Stress Responses: A Current View from the Points of Convergence in the Stress Signaling Networks," *Current Opinions in Plant Biology* 9 (2006), 436–42.

15 Xiong and Yang, "Disease Resistance."

16 J. Malamy, J. Hennig, and D. F. Klessig, "Temperature-Dependent Induction of Salicylic Acid and Its Conjugates During the Resistance Response to Tobacco Mosaic Virus Infection," *Plant Cell* 4 (1992), 359–66; see for review K. Bomblies, "Too Much of a Good Thing? Hybrid Necrosis as a By-product of Plant Immune System Diversification," *Botany* 87 (2009), 1013–22.

much is known about exactly how this happens, but one recent study has uncovered an intriguing clue, with implications for temperature adaptation. Scientists started with a widespread flowering weed called *Arabidopsis thaliana*. This species is widely used as a model for studying plant molecular biology, primarily because of its small, well-characterized genome and its ease of cultivation in a laboratory setting. The researchers studied a line carrying a hyperactive mutant version of an *R* protein that continuously sends signals as if the plant has been attacked. The resulting constant pathogen response makes the plants very sick, but only at low temperatures. The sick plants were subjected to further chemical mutagenesis. This process yielded several mutants that were still sick (in other words, the hyperactive defense response was still functioning) but no longer able to shut down pathogen responses when exposed to higher temperatures, as *Arabidopsis* normally would. The researchers found that all of these mutations occurred within the gene encoding the hyperactive *R* protein, indicating that in at least some cases the temperature responsiveness of *R*-protein-mediated immune activation might reside in the specific sequences of the *R* proteins, perhaps altering their stability, instead of in other proteins required for defense signaling.[17] This is encouraging, because it suggests that *R* genes from heat-tolerant species might provide a store of variants that can provide functional resistance at high temperatures to plants that

17 Y. Zhu, W. Qian, and J. Hua, "Temperature Modulates Plant Defense Responses Through NB-LRR Proteins," *PLoS Pathogens* 6 (2010), e1000844.

might not normally be able to respond to pathogens in hot weather or warmer climates. This also raises the interesting question of whether pathogen-signaling sensitivity to other stresses (e.g., salt, herbivore damage, humidity) could also be altered by changes in the R proteins.

Given that we can transfer functional resistance genes among species and that some R-protein variants can function at high temperatures, why not introduce more and more resistance genes into our crops from stress-tolerant species whose R proteins may be resistant to environmental shutdown? What is to stop us from exploiting the enormous pool of R-gene diversity to generate superplants that can fend off everything under a wide range of conditions? The raw materials may indeed be there; resistance to most pathogens and insect pests exists somewhere in the plant world, and so does the ability to respond at different temperatures.

Evolution provides a solid guide here, answering "Not so fast!" It turns out to be very risky to carry too many or excessively diverse R proteins; plants can thereby develop a form of autoimmunity. Certain combinations of naturally evolved R-protein variants can, in hybrid plants, result in interactions that trigger aberrant hyperactivation of immune responses.[18] This pathogen-response hyperactivation, though it does confer strong resistance to pathogens,

18 Bomblies, "Too Much of a Good Thing?"; K. Bomblies et al., "Autoimmune Response as a Mechanism for a Dobzhansky-Muller Type Incompatibility Syndrome in Plants," *PLoS Biology* 5 (2007), 1962–72; K. Bomblies and D. Weigel, "Hybrid Necrosis: Autoimmunity as a Potential Gene Flow Barrier for Plant Species," *Nature Reviews Genetics* 8 (2007), 382–9.

comes with a very heavy penalty. Plants with hyperactive immune systems are usually dwarfed, experience high levels of cell death, are often sterile, and frequently die prematurely. This constellation of symptoms, known as hybrid necrosis, is not uncommon in crosses among wild-plant species but is most prevalent in the progeny of crosses among strains of domesticated plants derived from breeding programs for increased resistance to pathogens.[19] This suggests that strong selection for increased resistance may select for *R*-gene variants that, while perhaps particularly effective, may also be prone to interacting with other proteins improperly and signaling even when no pathogen is present. Our understanding of the implications of such catastrophic interactions among natural variants of *R* proteins for large-scale patterns of plant immune-system evolution is in its earliest stages.

Returning to the question of adaptability of the plant immune system in the face of climate change, what can we conclude? Adaptability to novel pathogens may be quite high—at least in species that have not been heavily inbred—because of the staggering amount of diversity present in the *R*-gene repertoire. How this diversity is distributed in wild species remains largely unexplored, making it difficult to gauge the potential response to selection in these species. The frequent appearance of hybrid necrosis in breeding programs for strong resistance to pathogens and the cross talk among pathogen and stress-signaling pathways, however, warns us that unanticipated and potentially negative side effects

19 Bomblies et al., "Autoimmune Response."

may arise with strong selection for either pathogen or stress resistance.

The sensitivity of pathogen-response signaling to temperature and other stresses is a concern, even though the effects may vary from one species to the next. Can we use genetic engineering to rapidly help plants mount immune responses under a wider range of conditions? Since single mutations in R proteins may suffice to alter the temperature sensitivity of resistance, it is conceivable that variants can indeed be engineered to expand the range of conditions in which defense signaling can be activated. However, before altering the immune-system shutdown temperature, we need to consider the complex interactions among the plant's signaling systems and ask if there is any unacceptable cost to the plant in losing the ability to shut down responses at elevated temperatures. Natural interaction effects may conspire to cause plants with elevated pathogen-resistance ability to become unacceptably sensitive to other challenges or perhaps particularly prone to autoimmunity. These outcomes would need to be tested by bioengineering projects in each case.

For the time being, the bioengineering of R genes will be limited to a handful of pressing cases: for example, cases of recently invasive pathogens or areas where the environment is changing rapidly and the plants are highly susceptible. Altering all of the R proteins in a host is currently not feasible. Thus, promoting durable resistance to more than just individual pathogens in the face of elevated temperature will first require us to identify and understand the molecular basis of temperature adaptations more generally. The most fruitful avenue for understanding adaptation of plant immunity to elevated temperatures will probably be to investigate the solutions that plants have hit upon naturally, through evolution, to allow them to mount defenses in warmer conditions. The experiments to discover how they accomplished this are not especially difficult; the challenge is in choosing the best experimental systems. We need examples in genetically tractable species where populations or related species have adapted to contrasting environments. We can then use standard genetic mapping techniques to identify genetic variants associated with the ability to mount immune responses under stress. This information could then be used in breeding or improvement schemes to boost resilience of crop strains or to estimate the adaptive potential of wild plants.

Climate-change stresses facing plant species threaten our food security as well as many wild-plant species critical to ecosystems throughout the world. These stresses are also likely to alter the dynamics of the interactions between plants and pathogens. How, and how much, will be important to take into account in models designed to predict the effects of climate change on natural ecosystems and agricultural productivity. Studies must continue to improve our understanding of how environment affects important signaling pathways, such as disease responses; such studies will help us predict the outcomes of rapid environmental change and better anticipate potential problems. They will also allow us to assess inherent adaptive potential and develop ways to help plants weather these changes—for example, through informed conservation strategies, altered farming practices, and targeted breeding or transgenic approaches. These are already being widely employed in our major staple food crops to meet the challenge of increasing food production for the burgeoning human population in the face of increasingly variable climate.[20] Ideally, we would hope to leverage the combined powers of molecular and evolutionary biology to identify ways in which we can help plants become robustly resilient to environmental stresses while still remaining resistant to pathogens.

20 See, for example, D. B. Lobell et al., "Prioritizing Climate Change Adaptation Needs for Food Security in 2030," *Science* 319 (2008), 607–10; and S. H. Strauss, "Genomics, Genetic Engineering, and Domestication of Crops," *Science* 300 (2003), 61–2.

ASIF A. GHAZANFAR
is an associate professor in Princeton University's Neuroscience Institute and its Department of Psychology, with a joint appointment in the Department of Ecology and Evolutionary Biology. He holds a PhD in neurobiology from Duke University (1999). His laboratory studies the neurobiology and behavior of monkeys as a way of understanding the evolution and function of the human brain and human social behavior.

THE EMERGENCE OF HUMAN AUDIOVISUAL COMMUNICATION

ASIF A. GHAZANFAR

I think the brain is like an AM radio. In an AM radio transmission, a signal is generated at a specific frequency corresponding to a specific rate of amplitude modulation (AM). This AM signal carries information content, such as music or a talk show, and is broadcast from a local radio station via an antenna. To broadcast over greater distances, or to reduce interference from other signals, the power (but not the frequency) of the amplitude modulation can be increased. To receive this transmitted AM signal, your radio uses another antenna, but this antenna is nonspecific—it will capture any and all AM signals that impinge upon it. Your radio picks out a specific radio station by means of a tuner, which uses a principle called resonance to filter out unwanted signals by amplifying only that one frequency.

In the act of speaking—or communicating vocally, in the case of nonhuman primates—the speaker may be thought of as the radio station, broadcasting an AM speech signal via the mouth. This signal, along with all the noises in the background, is picked up by the antennalike ears of the listener. The listener's brain is pretuned to the specific AM frequency of the speech signal and thus amplifies it through resonance. Well, it's not actually a "specific frequency" (this is squishy biology, after all) but a narrow range between 3 and 8 Hz (cycles per second). The speech signal is amplitude-modulated (that is, it goes up and down in intensity) within a 3–8 Hz frequency that resonates with ongoing rhythms in the auditory regions of the listener's brain that are also 3–8 Hz in frequency.

To further amplify this vocal signal in noisy environments, humans evolved rhythmic facial movements with the same frequency to accompany speech. Together, these facial and vocal rhythms divide up, or "chunk," the speech rhythm so that the listener's brain can efficiently extract meaningful information from the signal. Vocal communication can thus be understood as emerging through interactions among the brain, body, and environment of the signaler and the body and brain of the receiver.

This "emergent communication" is not a new idea, but it has often been ignored. Neurobiologists typically think of communication among parts of the brain as taking place only through neuronal connections. However, neural states can cause or influence other neural states with the environment as a conduit. Our body movements cause changes in sensory input that in turn cause changes in neural activity, which then lead to other behaviors, and so forth. Behavior is a circular process, flowing through the nervous system into the muscles and reentering the nervous system through the sense organs; the brain is not isolated.

Take, for example, one side of a conversation. A speaker, with no awareness of doing so, adjusts the production of her speech patterns by monitoring how they sound in a given environment. The motor areas of her brain generate a command to say something and the body complies, producing an acoustic signal that blends with the ambient noise. The sound the speaker makes travels to her own ears and, through her auditory system, acts as a feedback signal, telling her motor system to adjust her vocal output if necessary. This is why you (and other primates) reflexively raise your voice in noisy environments (a response known as the Lombard effect) and why you speak too loudly while listening to your iPod—inaccurate feedback because your ears are plugged. Communication between the motor and auditory systems of the brain is coordinated in part through the vocal signal, with the air as a conduit. Let's now extend this idea to both sides of a conversation: the speaker and the listener.

THE BRAIN HAS RHYTHMS, AND SPEECH IS TUNED TO ONE OF THEM

The mammalian neocortex produces a set of rhythms. These are regular amplitude modulations, representing fluctuations between low and high excitability states; they range in frequency from very slow (0.05 Hz) to very fast (500 Hz). The various frequencies are thought to represent the size of the underlying neural network that produces them. Higher-frequency rhythms are limited to small, local networks of neurons, while low-frequency rhythms represent very large populations of neurons. The rhythms we know the most about are labeled with Greek letters, in the order in which they were discovered: delta (1–3 Hz), theta (3–8 Hz), alpha (8–14 Hz), beta (14–30 Hz), and gamma (>30 Hz). They are associated with different brain states, and under certain conditions multiple rhythms can occur simultaneously in a particular neocortical area or across neocortical areas. What is most remarkable is that these rhythms occur in every mammal ever investigated and in every neocortical area (in one context or another). They appear to be—in part, at least—the product of the uniform structure of the mammalian neocortex: every neocortical area, in every mammal, is a six-layered structure with roughly uniform input and output connections.

The speech signal—across all languages and contexts— has its own amplitude modulation, consisting of a rhythm that, as noted, ranges from 3 to 8 Hz. This is also the time scale for syllable production; that is, speakers produce syllables at roughly three to eight times per second. Recent theories of speech perception point out that the amplitude modulations in speech are well matched to at least one of those ubiquitous neocortical rhythms: the theta band, or the 3–8 Hz rhythm.[1] This suggests that the speech signal

---

1 D. Poeppel, "The Analysis of Speech in Different Temporal Integration Windows: Cerebral Lateralization as 'Asymmetric Sampling in Time,'" *Speech Communication* 41 (2003), 245–55; C. E. Schroeder et al., "Neuronal Oscillations and Visual Amplification of Speech," *Trends in Cognitive Science* 12 (2008), 106–13; C. Chandrasekaran et al., "The Natural Statistics of Audiovisual Speech," *PLoS Computational Biology* 5:e1000436 (2009).

resonates with (that is, amplifies) the ongoing theta rhythm in the auditory regions of the listener's brain, thereby increasing the signal-to-noise ratio and helping us to hear better. If this hypothesis has any validity, one would expect that if the speech signal is disrupted so that it no longer has a rhythm in that frequency range (akin to changing the radio station), speech perception should be disrupted. This is exactly what happens. When the amplitude modulations of speech are artificially sped up to exceed 8 Hz, the listener's auditory cortex fails to pick them up efficiently, and intelligibility drops to chance levels.

This resonance hypothesis also extends to the visual modality. Face-to-face human conversations are apprehended through both the visual and auditory modalities. In fact, audiovisual speech is the primary mode of speech perception, and it is well established that watching a speaker's face helps you hear better. In noisy situations—for example, at a cocktail party—watching a speaker's face is the equivalent of turning up the volume by 15 decibels. This amplification of speech by vision occurs, at least in part, because movements of the mouth during speech are closely correlated with the speech signal's amplitude modulations: when you speak, your mouth is opening and closing at a rate of 3–8 Hz, thus amplifying even more the neural signals in your listener's brain.

These findings suggest a kind of reverse way of thinking about the evolution of mammalian communication. We expect the properties of the brain to be tuned to the features salient in an animal's environment. Bats, for example, have an auditory cortex with neurons exquisitely tuned to the echolocation signals they emit and use to hunt down prey.

Humans and other primates have neuronal systems exquisitely sensitive to faces, because faces help mediate almost all our social interactions. The matching of the speaker's audiovisual speech rhythms to the listener's neocortical theta rhythm flips this standard evolutionary scenario on its head by suggesting that the structure of the brain itself (specifically the neocortex) shaped the evolution of vocal acoustics instead of the other way around. The basic patterns of neocortical anatomy that produce the set of fixed neural rhythms are, as noted, conserved throughout the mammalian lineage, and they pre-date the elaboration of vocal repertoires. Evolution apparently selected for vocalizations structured in such a way as to match one of those ongoing neural rhythms: the 3–8 Hz theta rhythm.

THE EVOLUTIONARY IMPLICATIONS

If that evolutionary hypothesis has any merit, one might expect to see the same rhythms in the vocalizations of species closely related to humans. Unfortunately, we have no recorded vocalizations of Neanderthals or any other protohumans, so we are left with a comparative approach. We look at existing species, such as various monkeys and apes, that are closely related to humans, and if we find they share a feature, we can infer that their last common ancestor also had that feature. In the present case, since the various neural frequency bands are independent of brain size (that is, they occur in all mammals), primates of various sizes should exhibit the same 3–8 Hz rhythm in their vocalizations that humans do. That seems to be the case. In the three species my laboratory has examined—the New

World cotton-top tamarin, the Old World rhesus monkey, and the chimpanzee—all vocalizations have a 3–8 Hz amplitude modulation. Thus, despite their radically different brain and body sizes (tamarins weigh little more than a pound, while chimps weigh well over a hundred pounds), they share with humans this vocal feature, suggesting that our last common ancestor also had vocalizations with a 3–8 Hz rhythm.

Humans and other primates share not only an acoustic rhythm but other aspects of vocal production as well. In all primates, vocalizations are produced by the coordinated movements of the lungs, larynx, and vocal tract (the pharynx, mouth, and nasal cavity). Movements of the vocal tract result in predictable deformations around the mouth and other parts of the face. These predictable visual cues can be combined with the auditory signal to enhance detection and discrimination in humans and other primate species. Monkeys, apes, and humans can match different vocal expressions to the appropriate facial postures; they can recognize the correspondence across the two sensory modalities. Thus, the visual signal can help primates hear and identify the auditory signal in noisy environments.

There is one puzzling difference, however, between monkeys and humans: although both produce an acoustic signal with a 3–8 Hz rhythm, only humans produce rhythmic mouth movements (with the same frequency) when they communicate with speech. So, for example, when you watch someone speaking, that person's mouth is opening and closing at a rate of roughly 3–8 Hz, the same rate as—and in sync with—the acoustic speech stream exiting his or her mouth. This is not true for other primates. Nonhuman primates produce vocalizations with a single ballistic mouth movement: they open their mouth once, a vocalization with a 3–8 Hz rhythm exits, and then the mouth closes. Because humans have a matching visual rhythm, their auditory signals get a bigger boost than those signals do in other primates. This suggests that the acoustic rhythm in humans, apes, and monkeys is likely produced by the same mechanism but that in humans the rhythmic visual component was added at a later time. So the question is, How did we humans evolve this rhythmic *audiovisual* speech from the rhythmic *auditory* speech present in monkeys?

One hypothesis is that the rhythm of audiovisual speech evolved through modification of preexisting cyclical jaw movements in ancestral primates.[2] For example, though rhythmic jaw movements are relatively rare during vocal production by nonhuman primates, they are extremely common as facial communicative gestures. Gestures such as the lip smacks and teeth grinds of macaques involve cyclical movements of the mouth and are not accompanied by vocalizations. In the course of human evolution, so the theory goes, these nonvocal rhythmic facial expressions were coupled with vocalizations. Of course, tests of such evolutionary hypotheses are difficult. Yet if rhythmic speech did evolve through the rhythmic facial expressions of ancestral primates, two predictions can be tested using the comparative approach. The first is that, like the visual

---

2 P. F. MacNeilage, "The Frame/Content Theory of Evolution of Speech Production," *Behavior and Brain Science* 21 (1998), 499–511.

part of audiovisual human speech, these rhythmic facial expressions in macaques should occur within the 3–8 Hz frequency range. The second is that neocortical structures sensitive to faces and audiovisual communication signals in humans and other primates should also be responsive to these rhythmic nonvocal facial expressions. Such sensitivity would indicate that the nonhuman primate brain was in some sense "prepared" for the evolution of audiovisual speech.

These predictions have been tested, and the data support the hypothesis.[3] The two rhythmic facial expressions of macaques—lip smacks and teeth grinds—were found to have a rhythm of 3–8 Hz, consistent with the idea that human audiovisual speech rhythms evolved from coupling vocalizations with rhythmic facial expressions. (Notably, not just any mouth movements in macaques fall into this frequency range: their chewing, for example, is consistently slower than their lip smacks and teeth grinds.) However, there would be no point to this evolved coupling of facial movements and voices if the brains of receivers weren't already able to handle this new type of communication signal. The superior temporal sulcus is one region of the human brain that is critical for audiovisual speech perception; monkeys have a homologous region in their brains that is important for their audiovisual communication signals. Remarkably, this region of the monkey's brain

---

3 A. A. Ghazanfar et al., "Dynamic, Rhythmic Facial Expressions and the Superior Temporal Sulcus of Macaque Monkeys: Implications for the Evolution of Audiovisual Speech," *European Journal of Neuroscience* 31 (2010), 1807–17.

is also sensitive to rhythmic facial expressions. This suggests that no new neocortical areas had to evolve in order to produce and perceive the newly evolved rhythmic audiovisual speech signals—making the evolutionary transition to rhythmic audiovisual communication easier.

THE BRAIN NEEDS INFORMATION IN CHUNKS

Just as a radio station uses the AM signal to get its message to someone's radio, primate signalers use amplitude-modulated vocal signals to access a listener's brain—and it appears that humans alone amplify this signal with a concomitant visual component. But getting a signal to its target is not the end of the transmission. In radio communication, once the signal has resonated with a properly tuned radio, a "demodulator" removes the amplitude modulation so that you can properly hear the meaningful content (a voice, music, etc.) in the signal. In the listener's brain, a similar process must be able to extract the meaningful content from the vocal signals. The brain does this not by demodulating the signal but by coupling the 3–8 Hz theta rhythm with a higher-frequency neocortical rhythm called gamma (>30 Hz). This gamma rhythm has the necessary temporal resolution to extract the finely structured acoustic content in a voice (e.g., vowels and consonants, as well as the acoustic content pertinent to the physical characteristics of the speaker, such as gender, size, and age).

At this point, you may be asking yourself, "Okay, a 3–8 Hz rhythm is important and couples with the gamma rhythm, but why? Why couldn't we just as well use the gamma rhythm alone?" The answer is that the 3–8 Hz rhythm in vocalization can "chunk" the signal into manageable units for the faster gamma rhythm. The auditory system is unique in that it cannot actively sample the sensory world. For example, in vision, we move our eyes here and there, focusing the high-resolution part of our eyes (the fovea) on features of interest. When we try to recognize objects by touch, we palpate them. When we smell things, we sniff. In vision, touch, and smell, we (and other animals) actively acquire sensory information—and, surprisingly, the rate of these movements is at roughly 3–8 Hz! The movements effectively chunk the incoming signals. The auditory system, on the other hand, has no active component. Yes, some animals can move their ears to better detect the location of a sound, but they generally don't flap their ears back and forth to actively sample and identify sound sources. Primates certainly don't. In the case of vocal communication, therefore, evolution imposed the chunking on the signaler's side of the equation by adding a 3–8 Hz amplitude modulation to vocalizations.

The evidence for this chunking solution in the auditory system extends from perceptual experiments with speech all the way down to single neurons. In one experiment, sentences were divided up into segments of fixed duration and then every other segment was played backward.[4] Human listeners presented with such disrupted sentences were nevertheless able to recognize the content of the sentence, but only when the reversed-segment durations were less

---

4 K. Saberi and D. R. Perrott, "Cognitive Restoration of Reversed Speech," *Nature* 398 (1999), 760.

than 100 milliseconds (ms), or a tenth of a second. When segment durations exceed 130 ms, intelligibility drops to chance levels. A 130 ms duration is a time interval that exceeds the wavelength of an 8 Hz rhythm—i.e., it is out of tune. Similarly, single neurons in the auditory regions of the macaque neocortex can discriminate between two vocalizations that differ only in that one indicates a small vocalizer and the other a large vocalizer, but when the 3–8 Hz rhythm is removed from the vocalizations, the neurons can no longer accomplish this simple discrimination. These and other findings suggest that the 3–8 Hz rhythm in vocalizations, working in concert with ongoing brain rhythms in the listener, is important for chunking the signal into manageable units for further neural processing and ultimately for perception.

Our brains, like AM radios, are parts of a larger system; they are inoperable without external components. This larger system involves a signaler that emits an amplitude-modulated signal, transmitted through the air and received by a listener's brain, that has at least one rhythm pretuned to the signal. In humans, this signal can be amplified using closely correlated facial movements. It seems, therefore, that the structure of vocalizations in primates exploits the structure of ongoing brain activity, with communication emerging as this interaction unfolds in time. The differences between human and nonhuman primates suggest that audiovisual speech may have evolved rather easily and without any radical changes to key brain structures or the development of new ones.

If communication is an emergent property of two people's rhythms interacting, it stands to reason that future work must use experimental paradigms that appropriately capture this interaction—the contingency between a signaler and a receiver. In our lab, we are moving toward having two monkeys interact with each other in a controlled environment, as well as having monkeys interact with monkey avatars.[5] In the latter case, we plan to control one side of the social interaction. These types of experiments should lead to a more realistic picture of what the brain is doing in real time and what may go awry in a variety of speech and communication disorders.

5 S. A. Steckenfinger and A. A. Ghazanfar, "Monkey Visual Behavior Falls into the Uncanny Valley," *Proceedings of the National Academy of Sciences* 106 (2009), 18362–6.

NAOMI I. EISENBERGER

has a BS in psychobiology and a PhD in social psychology (2005) from the University of California, Los Angeles. She is an assistant professor in UCLA's Social Psychology program, director of the Social and Affective Neuroscience Laboratory, and codirector of the Social Cognitive Neuroscience Laboratory. Her research focuses on using behavioral, physiological, and neuroimaging techniques to understand how the human need for social connection has left its mark on our minds, brains, and bodies.

WHY REJECTION HURTS

NAOMI I. EISENBERGER

"That *hurt* my feelings." "My heart was *broken*." If you listen closely to the ways in which people describe their experiences of social rejection, you will notice an interesting pattern: we use words representing physical pain to describe these psychologically distressing events. In fact, in the English language we have few means of expressing rejection-related feelings other than with words typically reserved for physical pain. Moreover, using such words to describe experiences of social rejection or exclusion is common to many languages and not unique to English.[1]

---

1 G. MacDonald and M. R. Leary, "Why Does Social Exclusion Hurt? The Relationship Between Social and Physical Pain," *Psychological Bulletin* 131 (2005), 202–23.

Why do we use words connoting physical pain to describe experiences of social rejection? Is feeling socially estranged truly comparable to feeling physical pain, or are these words to be regarded simply as figures of speech? My laboratory's research has suggested that the "pain" of social rejection ("social pain") may be more than just a figure of speech. Through a series of studies, my colleagues and I have shown that socially painful experiences, such as exclusion or rejection, are processed by some of the same neural regions that process physical pain. Here I review the evidence that led us to the notion that physical and social pain processes overlap and the studies that directly test this overlap. I will explore some of the potentially surprising consequences of such an overlap as well as what this shared neural circuitry means for our experience and understanding of social pain.

DOES REJECTION ACTUALLY HURT?

Though it may seem far-fetched to claim that social rejection can actually "hurt," an overlap in the distress associated with physical and social pain makes good sense from an evolutionary perspective. As a mammalian species, humans are born relatively immature, unable to feed or fend for themselves. Because of this, infants, in order to survive, must stay close to a caregiver to get the necessary nourishment and protection. Later on, connection to a social group becomes critical to survival; its members benefit from shared responsibility for gathering food, thwarting predators, and caring for offspring. Given that being socially disconnected is so detrimental to survival, it has been suggested that in the course of our evolutionary history the social attachment system—which ensures social closeness—piggybacked onto the physical-pain system, borrowing the pain signal to cue instances of social separation.[2] Social connection may have been so important for survival that the painful feelings associated with physical injury were co-opted to ensure that social separation was equally distressing—that individuals would be motivated by such feelings to avoid social disconnection and maintain closeness with others.

Research with animal and human subjects alike has indicated that physical and social pain processes overlap. Specifically, two brain regions—the dorsal anterior cingulate cortex (dACC) and to a lesser extent the anterior insula—seem to contribute both to the distress of physical pain and to behaviors indicative of separation distress in nonhuman mammals.

With regard to physical pain, the dACC and anterior insula seem to track the "affective" or unpleasant component of pain experience. The experience of pain can be divided into two components: the sensory component, which in part provides information about where the painful stimulus is felt, and the affective component, which registers the perceived unpleasantness of the stimulus—that is, how bothersome it is. Following neurosurgery to remove part of the dACC in order to relieve intractable chronic pain, patients report that they can still identify where painful stimuli are coming from but that the

---

2 J. Panksepp, *Affective Neuroscience* (New York: Oxford University Press, 1998), see especially chap. 14.

stimuli "no longer bother them."[3] Similar findings have been observed following damage to the anterior insula.[4] In contrast, damage to the somatosensory cortex, a region involved in pain localization, prevents patients from identifying where the pain is coming from but leaves the affective distress intact.[5] Neuroimaging studies also support this distinction. Subjects who were hypnotized to increase the "unpleasantness" of painful stimuli without altering the sensory component showed increased activity in the dACC but not in the primary somatosensory cortex, which supports the sensory component of pain.[6]

Interestingly, some of these same pain-related neural regions also contribute to specific behaviors associated with being separated from a caregiver—namely, distress vocalizations. Infants of many mammalian species emit distress vocalizations (for example, crying, in human infants) when separated from their caregivers. These serve the adaptive purpose of cueing the caregiver to retrieve the infant, thus preventing prolonged separation between the two. The ACC (both dorsal and ventral subdivisions) plays a critical role in producing these distress vocalizations.

3 E. L. Foltz and L. E. White, "Pain 'Relief' by Frontal Cingulumotomy," *Journal of Neurosurgery: Pediatrics*, 19 (1962), 89–100.

4 M. Berthier, S. Starkstein, and R. Leiguarda, "Behavioral Effects of Damage to the Right Insula and Surrounding Regions," *Cortex* 23 (1987), 673–8.

5 M. Ploner, H.-J. Freund, and A. Schnitzler, "Pain Affect Without Pain Sensation in a Patient with Postcentral Lesion," *Pain* 81 (1999), 211–14.

6 P. Rainville et al., "Pain Affect Encoded in Human Anterior Cingulate but Not Somatosensory Cortex," *Science* 277 (1977), 968–71.

Lesions to the ACC in squirrel monkeys eliminate distress vocalizations, whereas electrical stimulation of the ACC in rhesus monkeys leads to the spontaneous production of distress vocalizations.[7]

Based on these findings highlighting neural regions involved in both physical pain in humans and separation-distress behaviors in nonhuman mammals, we decided to investigate whether these regions would play a role in socially painful experiences in humans. In one such study, each participant was told that he or she would be connected over the Internet to two other individuals and that together they would be playing a computer game of catch while in the fMRI scanner. Through goggles, the participant saw cartoon representations of the two other players (along with their names), as well as his or her own hand, and with the press of a button the participant could decide which player to throw the ball to.[8]

In reality, there were no other players; the study's participants played with a preset computer program. In the first round of the game, participants were included for the entire

7 K. A. Hadland et al., "The Effect of Cingulate Lesions on Social Behaviour and Emotion," *Neuropsychology* 41 (2003), 919–31; P. D. MacLean and J. D. Newman, "Role of Midline Frontolimbic Cortex in Production of the Isolation Call of Squirrel Monkeys," *Brain Research* 45 (1988), 111–23; B. W. Robinson, "Neurological Aspects of Evoked Vocalizations," in *Social Communication Among Primates,* ed. S. A. Altmann (Chicago: University Press, 1967), 135–47; W. Smith, "The Functional Significance of the Rostral Cingular Cortex as Revealed by Its Responses to Electrical Excitation," *Journal of Neurophysiology* 8 (1945), 241–55.

8 N. I. Eisenberger, M. D. Lieberman, and K. D. Williams, "Does Rejection Hurt? An fMRI Study of Social Exclusion," *Science* 302 (2003), 290–2.

time, but in the second round they were socially excluded, when the two other (virtual) players stopped throwing the ball to them partway through the round. In response to this exclusion, the subjects showed significant activation in the dACC and anterior insula, two regions associated with the distress of physical pain. Moreover, subjects who reported feeling greater social distress in response to the exclusion episode ("I felt rejected," "I felt meaningless") also showed greater activity in the dACC, supporting the commonsense notion that rejection really does "hurt."[9]

Subsequent studies have supported these initial findings. Besides showing a relationship between in-the-moment reports of social distress and pain-related neural activity in response to social exclusion,[10] subjects who report feeling more rejected in their everyday social interactions show greater pain-related neural activity in response to an episode of social exclusion.[11] In some cases, simply viewing images of stimuli that signal social rejection triggers these pain-related neural regions. For example, viewing rejection-themed paintings, such as those by Edward Hopper, has been shown to activate the dACC and the anterior

9 N. I. Eisenberger and M. D. Lieberman, "Why Rejection Hurts: A Common Neural Alarm System for Physical and Social Pain," *Trends in Cognitive Science* 8 (2004), 294–300.

10 N. I. Eisenberger, S. L. Gable, and M. D. Lieberman, "fMRI Responses Relate to Differences in Real-World Social Experiences," *Emotion* 7 (2007), 745–54; K. Onoda et al., "Decreased Ventral Anterior Cingulate Cortex Activity Is Associated with Reduced Social Pain During Emotional Support," *Social Neuroscience* 4 (2009), 443–54.

11 Eisenberger, Gable, and Lieberman, "fMRI Responses."

insula.[12] Additionally, for rejection-sensitive individuals, viewing videos of individuals making disapproving facial expressions—a potential cue of social rejection—is associated with greater activity in the dACC.[13]

Finally, rejection and exclusion are not the only elicitors of pain-related neural activity. Other socially painful experiences, such as bereavement, appear to activate these neural regions as well. In response to viewing pictures of a recently deceased mother or sister (compared with a picture of a female stranger), female subjects showed increased activity in the dACC and anterior insula.[14] Moreover, females who lost an unborn child after induced termination, compared with those who delivered a healthy child, showed greater activity in the dACC in response to viewing pictures of smiling babies.[15] In sum, various kinds of socially painful experiences, from rejection to bereavement, seem to rely in part on neural regions that play a direct role in the experience of physical pain.

---

12 E. Kross et al., "Neural Dynamics of Rejection Sensitivity," *Journal of Cognitive Neuroscience* 19 (2007), 945–56.

13 L. J. Burklund, N. I. Eisenberger, and M. D. Lieberman, "Rejection Sensitivity Moderates Dorsal Anterior Cingulate Activity to Disapproving Facial Expressions," *Social Neuroscience* 2 (2007), 238–53.

14 H. Gündel et al., "Functional Neuroanatomy of Grief: An fMRI Study," *Journal of Psychiatry* 160 (2003), 1946–53; M. F. O'Connor et al., "Craving Love? Enduring Grief Activates Brain's Reward Center," *NeuroImage* 42 (2008), 969–72.

15 A. Kersting et al., "Neural Activation Underlying Acute Grief in Women After the Loss of an Unborn Child," *American Journal of Psychiatry* 166 (2009), 1402–10.

WHAT ARE THE CONSEQUENCES OF A
PHYSICAL-SOCIAL PAIN OVERLAP?

To the extent that physical and social pain processes overlap, one might expect some interesting consequences—for
example, that individuals who are more sensitive to physical pain would also be more sensitive to social pain, and
vice versa. Since this hypothesis is not an intuitive one, few
studies have directly investigated it. The best evidence for
it comes from findings from patient populations, showing,
for example, that adults with chronic pain are more likely
than healthy control subjects to worry about rejection by a
partner and that depressed patients with rejection sensitivity show greater pain sensitivity than do controls.[16]

To examine this possibility more directly, we investigated whether sensitivity to physical pain in healthy subjects was related to sensitivity to social rejection. In one
study, we demonstrated that participants who showed
greater sensitivity to heat-pain stimuli at baseline also
reported feeling more rejected by a subsequent experience
of social exclusion.[17] In a second study, we demonstrated
that individuals with the rare form of the mu-opioid recep-

---

16 P. Ciechanowski et al., "The Relationship of Attachment Style
to Depression, Catastrophizing and Health Care Utilization in
Patients with Chronic Pain," *Pain* 104 (2003), 627–37; A. Ehnvall
et al., "Pain During Depression and Relationship to Rejection Sensitivity," *Acta Psychiatrica Scandinavica* 119 (2009), 375–82.

17 N. I. Eisenberger et al., "An Experimental Study of Shared Sensitivity to Physical Pain and Social Rejection," *Pain* 126 (2006),
132–8.

tor gene (OPRM1), who are known from previous research to show greater sensitivity to physical pain, reported higher levels of rejection sensitivity and evidenced greater pain-related neural activity (dACC, anterior insula) in response to a scanner-based episode of social exclusion.[18]

A second consequence of a physical-social pain overlap is that factors that increase or decrease one kind of pain should affect the other kind of pain in a similar manner. Thus, factors typically thought to reduce social pain (such as feeling socially supported) should also reduce physical pain, and factors typically thought to reduce physical pain (such as pain medication) should also reduce social pain. Indeed, we have found evidence for both of these possibilities. To explore whether social support reduces physical pain, we asked female participants to rate the unpleasantness of a series of painful heat stimuli delivered to their forearm as they performed a number of different tasks.[19] In one of these tasks, participants received social support (e.g., they held their romantic relationship partner's hand), and in others they did not (e.g., they held a stranger's hand or a squeeze ball). We found that participants reported significantly less pain when holding their partner's hand than when they held a stranger's hand or a squeeze ball. What was more provocative, we also found that participants

18 B. M. Way, S. E. Taylor, and N. I. Eisenberger, "Variation in the Mu-opioid Receptor Gene (OPRM1) Is Associated with Dispositional and Neural Sensitivity to Social Rejection," *Proceedings of the National Academy of Sciences* 106 (2009), 15079–84.

19 S. L. Master et al., "A Picture's Worth: Partner Photographs Reduce Experimentally Induced Pain," *Psychological Science* 20 (2009), 1316–8.

reported feeling significantly less pain while simply viewing pictures of their partner than when viewing pictures of a stranger or an object. Apparently even reminders of one's social support figure may reduce physical pain as well as social pain.

When I give talks on this work, people often ask, "If all this is true, does that mean that painkillers could reduce the pain of social rejection?" The question is usually meant to be funny—since it seems so implausible—but in fact the answer is yes. To test the idea, we examined whether Tylenol could reduce sensitivity to social rejection.[20] In the first such study, participants took either a normal dose of Tylenol or a placebo for three weeks and were asked to report on their daily levels of "hurt feelings." Participants who took Tylenol reported, on average, a significant reduction in daily hurt feelings starting on day 9 and continuing through day 21, whereas participants who got the placebo reported no change. In a second study, a separate group took either Tylenol or a placebo each day for three weeks and then played the virtual ball-tossing game (in which they were eventually socially excluded) in the fMRI scanner. Consistent with the first study, participants who had been taking Tylenol showed significantly less pain-related neural activity (dACC, anterior insula) in response to being socially excluded. These studies indicate, perhaps surprisingly, that Tylenol, a common physical pain killer, can also reduce social pain.

Several other possible consequences of a physical-social

20 C. N. DeWall et al., "Tylenol Reduces Social Pain: Behavioral and Neural Evidence," *Psychological Science* 21 (2010), 931–7.

pain overlap have yet to be directly explored. One phenomenon that may be better understood in light of this overlap is rejection-induced aggression. For years, researchers have puzzled over extensive evidence showing that individuals who are socially rejected are more likely to act aggressively toward others. Indeed, some of these findings hit close to home, as in the well-publicized school shootings often carried out by perpetrators described as social outcasts. On its own, the finding that rejection promotes aggression makes little sense; after all, given the importance of maintaining social ties, why would one be predisposed to act aggressively rather than prosocially following social rejection? Wouldn't it be more adaptive to try to reestablish social connection following rejection? However, when considered in the light of a physical-social pain overlap, aggressive responses to social rejection make more sense. It is well known from animal research that when an animal receives painful stimulation, it will attack others nearby. This seems to serve an adaptive function: when one is in danger of being physically injured, one attacks. If the social-pain system does indeed co-opt parts of the physical-pain system, aggressive responses to social rejection might be a by-product of an adaptive response to physical pain—one that fails to serve an adaptive function in the context of social rejection.

Another possible consequence of the overlap may be found in the physiological stress responses that occur in socially threatening situations. It is well known that physically threatening situations induce physiological stress responses (e.g., increased cortisol levels) to mobilize energy and resources to deal with the threat. However, it has also been demonstrated that socially threatening situations—such as delivering a speech in front of an evaluative or rejecting audience—can result in similar physiological responses, such as increased cortisol levels.[21] Whereas it seems adaptive to mobilize energy resources to deal with a physically threatening situation, it seems unlikely that an organism would need to mobilize these energy resources to deal with the possibility of being evaluated negatively or rejected by others. However, if the threat of social rejection is interpreted by the brain in the same manner as the threat of physical harm, physiological stress responses might well be triggered in both situations.

One of the implications of these findings is that episodes of rejection or relationship dissolution can be just as damaging and debilitating as episodes of physical pain to the person experiencing them. Even though we may treat physical pain more seriously and regard it as the more valid ailment, the pain of social loss can be equally as distressing, as demonstrated by the activation of pain-related neural circuitry upon social disconnection.

One may wonder whether this physical-social pain overlap is an unfair burden that we, as humans, have to bear. I suggest that it is not. Although certainly painful in the short term, the misery and heartache following broken social relationships serve a valuable function—namely, to ensure the maintenance of close social ties. To the extent that being rejected hurts, individuals are motivated to

---

21 S. S. Dickerson and M. E. Kemeny, "Acute Stressors and Cortisol Responses: A Theoretical Integration and Synthesis of Laboratory Research," *Psychological Bulletin* 130 (2004), 355–91.

avoid situations in which rejection is likely. Over the course of evolutionary history, avoiding social rejection and remaining connected to others likely increased one's chances of staying alive and reproducing. The experience of social pain, while temporarily distressing and hurtful, is an evolutionary adaptation that promotes social bonding and, ultimately, survival.

JOSHUA KNOBE,
an assistant professor in the Philosophy Department and the Program in Cognitive Science at Yale University, is one of the founding members of the "experimental philosophy" movement. Much of his work proceeds by using the experimental methods associated with cognitive science to address the questions traditionally associated with philosophy. He received his PhD from Princeton University (2006) and has since published numerous papers in both psychology and philosophy.

FINDING THE MIND IN THE BODY

JOSHUA KNOBE

Suppose that you look out around you and see four things: a human being, a fish, a toaster, and a printer cartridge. Looking at those four things, you might be able to detect all sorts of similarities and differences, but there is one distinction that seems especially noticeable: some of those things have minds, while others do not. One would attribute a mind to the human being, perhaps also to the fish, but definitely not to the toaster or the printer cartridge.

This distinction comes so naturally to us that it is easy to take it for granted, but if you stop to think about it for a moment, it begins to seem deeply puzzling. How exactly do people decide which things have minds and which do not? Most of us don't have much background in experimental psychology, cognitive neuroscience, or any of the scientific disciplines that might be relevant to a question like this one. Yet somehow if we see a fish swimming in a pond and then see a toaster popping up some toast, we immediately have the intuition that the former might have certain psychological states but the latter most definitely does not. How might we be doing this?

One traditional answer says that the process is relatively straightforward: we figure out whether something has a mind by checking to see what it *does*. In this view, the key thing to notice about fish is that we can see them swimming around and responding to their environments in complex ways. Meanwhile, the toaster never seems to do anything interesting—all it ever does is make toast. So an obvious hypothesis would be that it is this difference in behavior that leads us to say that fish have minds while the toaster does not.

This hypothesis has a great deal of intuitive appeal. There is something that seems deeply right about the idea that we attribute minds to objects based on their behavior, and philosophers have developed complex conceptual frameworks that spell this idea out in sophisticated detail. Such frameworks typically say that our ordinary way of thinking about the mind is something like a scientific theory. Just as a physicist might make sense of scientific observations by positing unseen entities, people make sense of one another's behavior by positing unseen mental states. The only difference is that physicists explain their data in terms of purely physical factors (forces, particles, fields), whereas ordinary people explain behavior in terms of psychological factors (beliefs, intentions, emotions).

Yet although the intuitive appeal of this framework is

undeniable, it has recently come up against a challenge from a somewhat unexpected source. A group of people working in philosophy departments began thinking that it might be time to leave their armchairs and conduct some systematic experimental studies. These "experimental philosophers" were particularly interested in the traditional philosophical view that people's ordinary way of making sense of the world might be something like a scientific theory, and they therefore set out to put this claim to the test empirically. But, surprisingly, the experimental results did not end up conforming to the traditional view. Again and again, the results seemed to show that people's ordinary way of making sense of the world was radically different from anything we might expect to find in a purely scientific theory.

The issues here can be complex, but let us focus for the moment on just one aspect of the problem. People's intuitions about whether a given entity has a mind do not appear to be based entirely on a scientific attempt to explain that entity's behavior. Instead, these intuitions seem to be influenced in a striking way by questions about whether that entity has the right sort of *body*.

MIND WITHOUT BODY

If we want to get to the bottom of these questions, a natural place to start is by looking for an entity that has a humanlike pattern of behavior but doesn't have any kind of body. We need to find an entity that takes in information from the environment and uses this information in a complex way to attain goals. However, we also need to make sure that the entity does not have anything like a biological body in the familiar sense.

Fortunately, we can easily find an entity that meets these requirements: the modern corporation. There is a clear sense in which corporations have goals, take in information, and plan accordingly. But corporations do not have biological bodies. Instead of being made up of a head, torso, and limbs, they are made up of a complex hierarchy of departments and committees. So perhaps corporations can serve as a helpful case study for present purposes.

But now things begin to get interesting. The first thing to notice is that people do sometimes use sentences that seem to ascribe mental states to a corporation. For example, a person might say:

> Acme Corporation intends to release a new
>   product in July.

Now, one might initially suppose that sentences like that are just metaphors or loose talk and that people aren't actually thinking that these corporations can literally have anything like an intention. But the experimental evidence suggests that things might be a little bit more complex. In a recent neuroimaging study, I collaborated with the psychologists Adrianna Jenkins, David Dodell-Feder, and Rebecca Saxe to look at the patterns of brain activation observed when people read such sentences. It turns out that they show activation in the very same brain regions that have been traditionally associated with thinking about other minds, particularly the right temporoparietal junction, which seems to be highly selective for thinking about beliefs and intentions.

So suppose we assume for the moment that people actually do ascribe mental states to corporations. We now come to the crux of the issue: If people think that a corporation can have some kind of mind, what sort of mind do they think it can have? Do people conceive of a corporation as having the same sort of mind we might find in a human being or an animal, or do they think there is something important that a corporation is missing? To get at these questions, I teamed up with the philosopher Jesse Prinz and ran a series of studies. What we found was that people showed a systematic tendency to see corporations as having only one, highly delimited aspect of a normal mind. In particular, people were happy to agree with sentences like these:

> Acme Corporation believes that its profit margin
>   will soon increase.
> Acme Corporation intends to release a new
>   product this January.
> Acme Corporation wants to change its corporate
>   image.

But suppose we shift over to sentences that ascribe to a corporation some kind of feeling or experience. For example:

> Acme Corporation is now experiencing great joy.
> Acme Corporation is getting depressed.
> Acme Corporation is feeling upset.

People regarded these sentences as completely wrong, sometimes laughably so. In other words, people seem to think that corporations are capable of deciding, intending, knowing, and so on but that they are not capable of truly feeling or experiencing anything. In the jargon of philosophy, corporations are regarded as utterly lacking in *phenomenal consciousness*.

Here again, we seem to be faced with a result that initially seems perfectly natural and obvious but begins to look more and more puzzling as one examines it further. Why exactly can't a corporation feel upset? A corporation can certainly *behave* in ways that would be characteristic of feeling upset, yet one somehow gets the sense that the corporation wouldn't truly be *feeling* anything at all. How do we arrive at this intuition? One possible answer is that people don't think that corporations can have consciousness because corporations don't have the right sorts of bodies.

MINDS AND MACHINES

Of course, we will never be able to isolate the role of the body if we restrict our attention to corporations. Corporations differ from human beings in numerous respects, and any of these differences could explain the observed effects. What we really need is an entity that is almost exactly the same as a human being except that it lacks a biological body.

The philosopher Bryce Huebner came up with the perfect way to fulfill these requirements. He conducted a series of studies in which participants were asked to imagine a robot that has been designed to act like a person. Although the robot is presumably made of silicon and metal, it is described as behaving exactly like a human being on all possible psychological tests. The question now is what sorts of mental states people will be willing to ascribe to it.

Strikingly, the answer is that people ascribe to the robot exactly the same sorts of states that they are willing to ascribe to a corporation. They are happy to say:

> It believes that triangles have three sides.

But they are unwilling to say:

> It feels happy when it gets what it wants.

In other words, what we see arising here is exactly the same asymmetry we observed for intuitions about corporations. Once again, people are describing an entity without a biological body as having an ability to have states like beliefs but not as having a capacity for genuine feeling or experience.

But notice what is happening this time. The robot is described as behaving *exactly* like a human being in all situations. So any difference between the mental states we ascribe to it and the mental states we ascribe to a human being can't be understood in terms of an attempt to predict behavior. It must be that the body is playing some role here. Something about the presence of our faces, our flesh, our biological nature, must be triggering people to think that we have phenomenal consciousness.

MIND AND FLESH

Thus far, we have been considering cases in which an entity is seen as not having a body. But suppose we now go in the opposite direction. Suppose we find a case in which an entity is seen as *especially* embodied—a case in which we associate the entity with a body even more than we would in cases of ordinary human interaction. What sort of mind would they ascribe in a case like that?

I was discussing this question one day with the psychologist Kurt Gray when someone happened to overhear us and suggested an interesting new approach. It turned out that there was a recently published photography book in which each model was depicted in different stages of undress. This fact alone may seem a bit unsurprising, but in this particular case the book was composed in a particularly systematic way. For each of the models, there was a completely clothed photograph and then, on the facing page, a photograph of the same person, in the same position, with the same facial expression, only this time completely naked. It was a cognitive scientist's dream, the artistic equivalent of a perfectly controlled study. We immediately decided to run a new experiment with the pictures from this book as stimulus materials.

In collaboration with our colleagues Paul Bloom, Mark Sheskin, and Lisa Feldman Barrett, we put together the experimental design. Each participant would receive a photograph and then would be asked to guess, just on the basis of that one photograph, how much the model depicted was capable of having various different mental states. So a participant might be asked:

> Compared to the average person, how much is
>   Erin capable of self-control?
> Compared to the average person, how much
>   is Erin capable of feeling fear?
> Compared to the average person, how much is
>   Erin capable of planning?

But the stimuli were designed in such a way as to let us look systematically at the relationships between different variables. The questions asked both about ordinary nonphenomenal states (self-control, planning) and about states that involved phenomenal consciousness (feeling fear, feeling pleasure). Each participant was randomly assigned to receive either a picture of a clothed model or a picture of a naked model.

Before I tell you the results, let's take a moment to consider the predictions one might make in a case like this. One obvious prediction would be that showing the model naked would make her vulnerable to a kind of "objectification." Participants might come to think of her more as a physical object, a mere thing, and they might therefore be less inclined to see her as having mental states. But then we can also imagine another sort of prediction, going in a quite different direction. Perhaps there is something about an awareness of the body that makes people more inclined to ascribe phenomenal consciousness. So the effect might actually end up going the opposite way. A focus on the body could make participants *more* inclined to think in terms of feelings and experiences.

With these predictions in mind, we can now turn to the actual results. For the nonphenomenal states, we ended up finding exactly what one would expect. The more salient a person's body was made, the less inclined participants were to ascribe those states. In other words, if you want people to take you as someone capable of complex planning and self-control, your best bet is not to have them looking at pictures of you naked. No surprises there. That is exactly the point that has been made repeatedly, and with great sophistication, in existing work within feminist theory.

But now comes the surprising part. For the phenomenal states, we did not find this same pattern. In fact, we found just the opposite. When the model was depicted naked, people were actually more inclined to think she was capable of having feelings and experiences. They were more inclined to think she was capable of feeling fear, more inclined to think she was capable of feeling pleasure. In fact, on all the different measures we used, we found that making the body more salient made people more inclined to ascribe feelings.

But the effect does not stop there. Gray has tested this same basic hypothesis using a whole series of imaginative experimental techniques. He has given participants information about a person's blood type, asked them to judge a person's physical attractiveness, even shown them pornographic images. Always, the result is the same. The more one makes participants focus on the body, the more they tend to ascribe feelings and experiences.

All in all, then, it does not appear that these phenomena are best understood in terms of a notion of "objectification." It is not as though participants are coming to think of a person as being a mere *object,* like a toaster or a printer cartridge. Rather, what we see emerging is a more complex pattern. Participants are thinking of the person as having less of one part of the mind but more of another. So perhaps it would be better to say that a focus on the body leaves us thinking of the person as an *animal.* That is, it leads us to think of the person as having more of the part of the mind we associate with animals (fear, pleasure, pain) and less of the part we regard as distinctively human (complex reasoning, planning, self-control).

PERCEIVING THE MIND

With all this experimental evidence on the table, we can now return to our original question. We wanted to get a better understanding of the process people use to figure out which entities have minds and which do not. So what exactly is this experimental evidence telling us?

The answer may come as a surprise. The key message coming out of the experimental evidence seems to be that the whole question was a mistaken one. It is beginning to look as though we might have been wrong to go searching for something such as "the process people use to figure out which entities have minds." The trouble is that there doesn't seem to be any single unified process that fits the bill. Instead, there appear to be two distinct processes here—one for figuring out whether an entity is capable of having states such as beliefs and goals, another for figuring out whether it is capable of genuine feelings. If we really want to get to the bottom of these issues, we will have to address each of these processes separately.

But when the question is reformulated in this way, a new answer becomes possible. We might discover that people's understanding of beliefs and goals involves something like a quasi-scientific attempt to explain human behavior, but we should not immediately assume that the same holds for people's understanding of feelings and experiences. On the contrary, all the evidence suggests that this latter process is deeply different. Our ordinary attributions of phenomenal consciousness do not appear to be based entirely on behavior; they do not appear to be purely scientific; they do not seem to serve primarily to aid prediction or explanation. Above all, they seem to be wrapped up in some fundamental way with an awareness of the body.

FIERY CUSHMAN

studies moral psychology. His research focuses on the cognitive processes that give rise to moral judgment, their development, and their evolutionary history. Dr. Cushman received a BA in biology from Harvard College in 2003 and a PhD in psychology from Harvard University in 2008. He is currently an assistant professor in the Department of Cognitive, Linguistic, and Psychological Sciences at Brown University.

SHOULD THE LAW DEPEND ON LUCK?

FIERY CUSHMAN

On a snowy Sunday afternoon, Hal and Peter watch football and drink beers at a local bar. Both drive away intoxicated, and both lose control of their cars on the slick roads. Hal collides with a neighbor's tree, while Peter collides with a young girl playing in the snow and kills her. In my home state of Massachusetts, Hal can expect a fine of several hundred dollars, along with temporary suspension of his license. Peter's punishment is far different: for vehicular manslaughter, he faces between two and a half and fifteen years in prison.

Cases like this have long vexed philosophers and legal scholars, who refer to the problem as "moral luck." Should a chance outcome exert such a powerful influence on our moral judgments? Such cases are troubling because we are caught between two irreconcilable perspectives. It seems wrong to punish Hal and Peter differently when they engaged in absolutely identical behavior. Yet it doesn't seem right to send Hal to prison for years for having hit a tree, even while driving under the influence; neither does it seem right to let Peter off with a fine for killing someone.

A growing body of psychological research shows that we feel conflicted about cases like Hal's and Peter's because we are literally of two minds. One mental mechanism assigns punishment in proportion to the harm that a person causes and therefore judges Hal and Peter very differently. Another mental mechanism judges actions to be morally wrong based on the intent to harm, or the risk of harm, and therefore judges Hal and Peter identically. When the divergent outputs of these mental mechanisms are compared, the result is a difficult moral dilemma. Insights into the psychological origins of "moral luck" add a new dimension to the philosophical problem: How will advances in the science of moral judgment change the way we think about the law?

In the 1930s, a young Swiss psychologist named Jean Piaget began to ask young children simple questions about right and wrong and thus gave birth to the field of moral psychology. By the time of his death in 1980, Piaget had become the most influential developmental psychologist in history, having contributed countless groundbreaking findings about the thought processes of young children. Piaget's great gift was to find cases in which children make judgments that are strikingly different from those of adults, and his study of moral judgment is no exception.

In a famous test of moral judgment, Piaget told children

stories about two young boys who broke teacups. One boy was trying to set the table to help his mother but dropped a tray of teacups and broke fifteen of them. The other boy was trying to steal cookies while his mother wasn't looking, and he knocked a single teacup to the floor. Piaget asked the children which of the boys was naughtier. To an adult, the answer is perfectly obvious. The boy who was trying to steal cookies was naughtier, because breaking the teacup was a result of his bad intention to steal, whereas the boy who broke fifteen teacups acted with good intentions. But Piaget found that the opposite answer was the obvious one to most children under six. The child who broke fifteen teacups was naughtier because fifteen teacups are a lot more than one.

Piaget's experiment reveals a basic tension between two factors that shape our moral judgment: the harm a person causes and the harm a person intends. Over the past eighty years, his experiment has been replicated and extended dozens of times, with consistent results. Young children tend to focus on causal responsibility, while older children and adults tend to focus on intent.

A few of these replications hint at an interesting asymmetry in the contest between causation and intention. When a person intends to do harm but doesn't (for instance, throwing a ball at somebody's face but missing), even very young children recognize that the perpetrator was naughty; in this and similar instances, young children are perfectly able to see that intentions matter. What appears to give children trouble is the opposite situation: when a person causes harm without intending to (for instance, aiming the ball at a target but accidentally hitting somebody in the face). To follow up on these suggestive findings, I sent a team of undergraduate research assistants to Boston's Museum of Science to interview hundreds of preschoolers. Our results confirmed this asymmetry: young children easily condemn people with bad intentions even when they don't cause harm, but they have trouble using good intentions to excuse somebody who does cause harm. In short, kids are tough on accidents.

Our study added another twist to Piaget's by asking children seven and younger to make two different kinds of moral judgments: about naughtiness and punishment. Here again, we found an interesting asymmetry. Most of the seven-year-olds reacted like adults, refusing to say that somebody was a "bad, naughty boy" if the harm he caused was accidental. Not so with punishment: a substantial fraction of this group claimed that accidental harm-doers should be punished. If kids are tough on accidents, they are really tough on punishing accidents. In that way, their judgments match our laws. (Remember the extra years in prison Peter got for causing a death, though his intentions were no different from Hal's.)

Is it possible that even adults would punish an accident? To put that question to the test, I conducted an online survey of more than 1,000 adults. They were asked to read a number of hypothetical scenarios and make a moral judgment after each one. Some were asked to make judgments about "moral wrongness," and others were asked to decide the "punishment deserved." The pattern of adult judgments was strikingly similar to that of our seven-year-olds—that is, judgments of moral wrongness depended almost exclusively on intent. An attempt to harm was judged very wrong whether or not it succeeded, while accidents were fully excused. But judgments regarding punishment were strongly influenced by causal responsibility. Attempts to harm were punished more severely when they succeeded than when they failed, and accidental harm was not fully excused.

This asymmetry between judgments about wrongness and judgments about punishment can explain why moral luck presents a dilemma. From the perspective of wrongness, Hal and Peter look the same: both intentionally drove under the influence of alcohol. But from the perspective of punishment, their actions produced very different outcomes: Hal caused negligible harm, whereas Peter ended a young girl's life. Either way we try to settle the case of Peter and Hal, a part of our brain is dissatisfied.

Over the years, philosophers and legal scholars have devised many creative, complex justifications for moral luck. But the psychology of the phenomenon seems to be in place among preschool children. When a harmful outcome occurs, it triggers a strong moral judgment. Very young children find it especially hard to override this negative judgment, even if the harm was accidental. By the age of seven, they tend to say that causing accidental harm is not morally wrong. But even in adulthood, the tendency to punish harmful accidents persists. To be sure, the different sentences that our laws hand down to Peter and Hal probably depend in part on sophisticated legal principles and policy considerations. But they probably also depend on a basic psychological impulse present in the youngest minds: the impulse to punish those who cause harm.

When people say that somebody should be punished for

an accident, do they really mean it? Perhaps it's just cheap talk. Perhaps they're trying to sound "tough on crime," or maybe they don't believe that the harm—as described in court, say, or in my scenarios—is truly accidental. Our next study of punishment addressed these concerns by designing a game that created actual accidents in the lab. We made sure that the accidents were just that—perfectly unintentional—and then we gave our subjects the chance to reward or punish those accidents with real money.

At the heart of the game was a simple division of money: Player A was given $10 to divide between herself and Player B. She could keep all $10 (stingy), divide it $5/$5 (fair), or give all $10 to Player B (generous). In order to introduce accidents, we built in a catch: Player A had to make this division by choosing to roll one of three different dice. Die 1 was stingy if it came up 1, 2, 3, or 4—but a 5 was fair and a 6 was generous. So, if you wanted to be stingy, you would roll die 1, though you might accidentally end up fair or generous by rolling a 5 or a 6. Dice 2 and 3 worked similarly: you could try to be fair by rolling die 2, or you could try to be generous by rolling die 3, but each of these dice, too, had a small chance of producing an accidental outcome.

By noting the die that Player A selected, Player B could know exactly what A's intentions were; even so, we asked B several questions to make sure he understood A's intent. We gave Player B the chance to reward or punish Player A by adding money to or subtracting money from Player A's payoff once she had rolled her die. When accidents occurred—for example, when A rolled the "generous" die but it came up 6 (stingy)—would Player B choose to reward or punish her? That is, would he focus on her intentions (the die she chose) or on the outcome (the way the die came up)?

The results indicated a strong role for accidental outcomes. When Player A chose the stingy die but it came up generous, on average Player B responded by rewarding her. And when Player A chose the generous die but it came up stingy, on average Player B responded by punishing her. Statistical analysis showed that Player B paid attention both to Player A's intentions and to the outcome of the roll but that outcomes mattered slightly more. It's behavior like this that puts the "luck" into moral luck. Our impulse to punish somebody who causes harm to us sometimes depends on nothing more than a roll of the dice.

This automatic search for causal responsibility in the face of harm can lead to unpredictable, even bizarre patterns of punishment. One example comes from a study I conducted on people's judgments of attempted murder. First I asked one group of participants to say how severely a competitive runner should be punished for having tried to kill a rival by sprinkling his salad with poppy seeds, to which the rival was supposedly fatally allergic. Fortunately, the fatal allergy turned out to be to hazelnuts, not poppy seeds, so the rival was just fine. Most people thought the runner should be punished for attempted murder, assigning an average sentence of about twenty years in prison.

Next I asked another group to judge a slightly different case. Once again, the runner sprinkled poppy seeds on his rival's salad, and once again the rival was actually allergic to hazelnuts. But this time, the rival's salad happened to have been made with hazelnuts by the chef. Consequently—and completely coincidentally—the rival died. From a rational perspective, this coincidence shouldn't affect the runner's punishment at all: he still performed exactly the same attempted murder. Yet people were nearly twice as likely to let him off the hook when the rival died coincidentally. I've now tested dozens of cases like this with hundreds of people, and the results are strikingly consistent: attempted crimes are more often excused when a harm occurs coincidentally, compared with attempted crimes in which the harm doesn't occur at all.

The reason is our basic impulse to assign causal responsibility for harm. When the rival dies from eating the salad, who is causally responsible? The chef, not the runner. People get focused on causal responsibility and assign less importance to the failed attempt at murder. If, however, the rival isn't harmed at all, no assessment of causal responsibility is required, and this allows people to fully consider the runner's murderous intent.

All of these studies point to the same conclusion about moral luck. When it comes to punishment, people care a lot about causal responsibility: when you cause harm, you tend to get punished, even if the harm was accidental. And when you *don't* cause harm, you might be let off the hook, even if you were attempting to murder. But why is this so? And what should we do about it?

It's impossible to say with certainty why our brain is designed the way it is. The forces of evolution shaped our brain over many millions of years, and we can't turn back the clock. Moreover, many of the thoughts we think weren't designed by evolution at all—our thoughts about quantum physics, say, or global warming. But for many patterns of human thought, we can make a good guess about evolutionary design. For instance, sexual intercourse between siblings is forbidden by most human cultures throughout the world and across history. Most animals avoid incest, and even many plants have structures designed to avoid self-pollination. Incest has very clear genetic costs, and that means loss of fitness in the evolutionary struggle for survival. It's a good bet that evolution designed our brains to avoid sex with siblings.

It's also likely that evolution has a hand in our instinct to punish. Biologists and economists have used mathematical models and creative experiments to show that punishment can play an important role in supporting good social behavior. This conclusion is unsurprising: we punish in order to teach people how to behave. But how can we explain the tendency to punish accidents? Is there some advantage to punishing a person who caused harm unintentionally?

In the end, the answer to this question depends on how people *learn* from punishment. To see this point more clearly, let's return to Piaget's example of the boy who breaks fifteen teacups when he tries to help his mother set the table. How should his mother react? From one perspective, she should punish him, as a means of teaching him not to break any more cups. But from another perspective, she should reward him, as a means of teaching him to continue trying to help others. Which is the best strategy for Mom? It all depends on the way her son learns. If he assumes that punishments and rewards relate to the outcomes he causes, punishment will teach him not to break cups. But if he assumes that punishments and rewards relate to the outcomes he intends, rewards will teach him to help set tables. Mom's best move depends on the design of her son's brain.

In order to answer this basic question about human learning, we set up another accident-prone game in the laboratory, this one involving darts instead of dice. Player A throws darts at a board with four colors on it, and, depending on what colors she hits, she wins or loses money for Player B. Here's the catch: Player A doesn't know which colors are good for Player B (earning him money) and which colors are bad (costing him money). The only way Player B can teach Player A is to reward or punish her after each throw. If Player B uses his punishments and rewards correctly, he can teach Player A to aim for the most valuable colors over the course of a dozen throws.

Critically, Player A has to call her shots, announcing which color she's aiming for. (The experimenter rewards her for accurate throws, thus ensuring her honesty.) Because our subjects weren't adept at darts, about half the time they hit the wrong color. This set up a choice for Player B: Should he reward or punish Player A for the color she aimed at, or should he reward or punish Player A for the color she hit? In order to find out, we secretly instructed Player B to reward and punish based on *hits* for half our participants, and to reward and punish based on *aims* for the other half. Then we measured which strategy did the better job of teaching A which colors Player B preferred.

The results were clear: Player A learned which colors were good for Player B about twice as effectively when rewarded and punished based on what she hit. Punishing accidental outcomes was the best strategy in this game, because it turns out that people do indeed learn from accidental outcomes. Although we cannot know for sure, this suggests a possible evolutionary explanation for why we tend to punish bad outcomes. The dart game re-creates a time in our evolutionary past when the only way to say "Do this!" was to reward and the only way to say "Don't do that!" was to punish. Given those circumstances, the results suggest that punishing accidents is a good idea. Maybe you stepped on my tail by accident, but when I bite you, you'll learn to watch out next time!

Yet, as with many of our evolutionary instincts, it's fair to question whether this one is still advantageous. For instance, we evolved an insatiable appetite for fat and sugar when they were scarce nutrients, but today over-consumption of those foods leads to heart disease and diabetes. Punishing accidents may have been the best way to teach good behavior in a time before we had language and laws, but in the modern world we can speak much more effectively with our tongues than our fists. The tea-cup Mom doesn't need to punish her son in order to say "Don't break teacups"—she can just tell him. And if I had not prohibited the players in the dart game from talking to each other, surely Player B would have said to Player A, "I lose money when you hit red, so I'm going to punish you if you even aim for it." Because we can use language and laws to communicate moral rules before they're violated, we may have outgrown the evolutionary necessity of punishing accidents.

By the time we're seven years old, we know that Hal and Peter acted equally wrongly when they drove drunk. For their reckless behavior, surely each deserves to be punished—but why not punish them with equal severity? In our evolutionary past, extra punishment for Peter was the best way to communicate a moral rule. In our evolutionary present, maybe we can let our laws do that work.

LIANE YOUNG

received her BA in philosophy (2004) and her PhD in psychology (2008) from Harvard University, after which she did postdoctoral work in MIT's Brain and Cognitive Sciences Department. She is an assistant professor in the Department of Psychology at Boston College, where she studies the cognitive and neural bases of human moral judgment. Her current research focuses on the role of reasoning and emotions in moral judgment and behavior—employing the tools and methods of social psychology and cognitive neuroscience, including fMRI (functional magnetic resonance imaging), TMS (transcranial magnetic stimulation), and the study of populations of patients with cognitive and neural deficits.

HOW WE READ PEOPLE'S MORAL MINDS

LIANE YOUNG

Every day, we observe and experience the effects of other people's actions. Often such observation and experience lead us to evaluate the actors themselves, especially in cases where their actions are harmful or helpful. Yet the harmful or helpful outcomes aren't all that matter. What matters more to us, in fact, is whether those outcomes were intended. We want to know what people were thinking when they acted. Did he know that selecting that option would erase my hard drive? Did she intend to dye my hair purple? Figuring out what's going on in someone's mind presents a challenge, to be sure, but a worthy one if we're to decide whether to forgive or condemn and how much—and who deserves our trust and friendship and who doesn't. How can we meet this challenge?

Investigating what goes on in the mind and brain when we make moral judgments can help unpack a simple rule we apply across multiple contexts, as participants in social relationships, as jurors in a courtroom, as parents or teachers of young children: it is morally wrong to intend harm. Recent work suggests that our moral judgment of another person depends on specific brain regions for reasoning about that other person's mental state (such as the person's belief or intent) as well as on other brain regions for generating emotional responses to such mental-state content. Studies reveal robust individual differences among people in their assignments of forgiveness and blame—differences that correlate with differences in their brain activity. Other studies show impaired moral judgments in patients with neurodevelopmental disorders or brain damage as well as changes in people's moral judgments when their brain activity is experimentally modulated.

Let's start with the role of a perpetrator's mental states in our judgment of the perpetrator. Consider the following scenario: Grace and her coworker are taking a tour of a chemical factory. Grace stops to pour herself and her coworker some coffee. Nearby, there's a container of sugar. The container, however, has been mislabeled "toxic," so Grace thinks that the powder inside is toxic. She spoons some into her coworker's coffee and takes none for herself. Her coworker drinks the coffee—and, of course, nothing bad happens. When experimental participants are presented with such a scenario, most say that what Grace did was seriously morally wrong and that Grace is seriously morally blameworthy, simply on the basis of her harmful intent.[1]

Participants also considered this alternative scenario. Near the coffee machine is a container of poison. The container has been mislabeled "sugar," so Grace thinks the powder inside is sugar. She spoons some into her coworker's coffee. Her coworker drinks it and falls down dead. Again, most participants judged Grace on the basis of her mental state. That is, they were prone to let Grace off the hook because of her false belief and her innocent intention.

In sum, we judge failed attempts at harm to be morally forbidden and accidental harm to be more or less permissible. We judge actions to be morally wrong when there is intent to harm, regardless of whether harm is done. When there is no intent to harm—when harm is done but by accident—we tend to be lenient. These behavioral patterns reflect the importance of reasoning about people's beliefs and intentions in evaluating their actions in moral terms.

MORAL LUCK

There is still, however, a significant difference in our moral judgments of accidents versus fully neutral acts (i.e., acts performed with neutral beliefs and intentions and resulting in neutral outcomes). People judge accidents as *somewhat* morally forbidden and assign *some* moral blame to the

---

1 L. Young et al., "The Neural Basis of the Interaction Between Theory of Mind and Moral Judgment," *Proceedings of the National Academy of Sciences* 104 (2007), 8235–40.

responsible agents. Grace is judged morally worse when she accidentally poisons her coworker than in the fully neutral situation, in which the container marked "sugar" actually contains sugar—even though Grace's intent was the same in the two situations.

What differs appears to be a matter of luck: a lucky (good) outcome (e.g., sweetened coffee) versus an unlucky (bad) outcome (e.g., poisoned coffee). Many moral judgments show this "moral luck" asymmetry: unlucky agents, who cause harm by accident, are usually seen as morally worse. Yet it also seems to most of us that moral matters ought not to be matters of luck. Is the case of accidental harm an exception to the rule that beliefs and intentions, not lucky or unlucky outcomes, determine an agent's moral status? What accounts for the different moral judgments we assign to lucky versus unlucky agents?

There are two candidate factors. For instance, unlucky Grace not only causes a bad outcome (poisoning her coworker) but holds a false belief (that the powder is sugar). Lucky Grace not only doesn't cause a bad outcome but holds a true belief (that the powder is sugar). These dimensions (belief and outcome) have typically been confounded in investigations of moral luck, making it impossible to tell whether the asymmetry in moral judgments we recognize as moral luck is due to the difference between lucky (good) and unlucky (bad) outcomes or the difference between true and false beliefs.

To test for the relative contributions of beliefs and outcomes to moral judgments, my colleagues and I developed a new scenario, featuring "extra-lucky" agents—that is, agents who hold the same false beliefs as unlucky agents but, thanks to an extra stroke of luck, don't cause any harm. Extra-lucky Grace falsely believes that the powder is sugar when in fact it is poison. She spoons the poison into her coworker's coffee. However, her coworker puts the coffee down and forgets to drink it, so no harm occurs. We hypothesized that extra-lucky Grace here would still be judged morally blameworthy on the basis of her false belief, even in the absence of any harmful outcome. Indeed, just as we hypothesized, extra-lucky Grace was judged more like unlucky Grace (who held the same false belief) than lucky Grace (who caused the same neutral outcome). The difference between true and false beliefs mattered more for moral judgments than the difference between neutral and bad outcomes. Furthermore, we found that the moral judgments made by our participants were critically affected by their assessments of whether the agent was justified in holding a belief—for example, that the toxic powder was actually sugar.[2] False beliefs were judged to be somewhat unjustified; therefore, agents holding those beliefs were judged to be somewhat blameworthy. Experiments like this reveal that mental-state factors (e.g., the truth and justification of an agent's beliefs) matter when we're making moral judgments, even when we're blaming people for accidents.

THE NEURAL BASIS OF REPRESENTING MENTAL STATES FOR MORALITY

Recent work has targeted the neural basis of our ability to reason about people's mental states when we make moral

---

2 L. Young, S. Nichols, and R. Saxe, "Investigating the Neural and Cognitive Basis of Moral Luck: It's Not What You Do but What You Know," *Review of Philosophy and Psychology* (in press).

judgments. This work builds on earlier work identifying specific brain regions for reasoning about mental states in nonmoral contexts—for instance, when we need to predict or interpret another person's behavior. Much of this earlier work uses a task from developmental psychology—the false-belief task, designed to test mental state reasoning in young children. In one scenario, children watch as a character named Sally places a ball in a basket and exits the room. Another character, named Anne, enters and moves Sally's ball to a box. Sally then returns, and the children are asked where Sally thinks her ball is. Children under four generally go for the box, because they cannot represent Sally's mental state as distinct from the real state of the world. Older children and adults pass the test by doing just that—representing Sally's false belief.

The false-belief task has been used to identify brain regions that support mental state reasoning. A number of fMRI studies show that a group of brain regions is selectively recruited by the false-belief task, compared with the brain regions recruited for a "control" task, such as reasoning about where to locate objects in outdated photographs or maps, or other nonmental representations. Research by Rebecca Saxe and her colleagues suggests that one of those brain regions—the right temporo-parietal junction (RTPJ), a patch of cortex above and behind the right ear—is selectively active in processing information about people's mental states as opposed to other kinds of information.

How does the RTPJ help us reason about mental states when we make moral judgments? In recent work, my colleagues and I conducted brain scans of participants while they read moral scenarios, such as the ones featuring Grace and her coffee mishaps. In one experiment, we varied the order in which we presented Grace's beliefs and facts about her action's outcome. For example, in half the trials, participants would read first that the powder was sugar and then that Grace believed the powder was poison. We reversed this order for the other trials: participants read about belief information first and facts about the outcome second. What we found was that the magnitude of the response in the RTPJ depended on whether belief or outcome information was being presented; the neural response was selectively higher for information about beliefs than for information about outcomes.[3] This result suggests that the neural response is sensitive to the presence of explicit mental states such as beliefs and that the RTPJ in particular supports the encoding of beliefs that are relevant for moral judgment.

Is there a relationship between this neural response and the moral judgment made? Does higher activity in the RTPJ predict greater reliance on mental states for making moral judgments? When we scanned the brains of undergraduate participants, we observed individual differences both in the moral judgments they made and in the magnitude of the neural responses when they made the judgments. Some participants judged accidental harm as very blameworthy, while other participants judged accidental harm as not very blameworthy at all. More important, these differences correlated with differences in the neural

---

3 L. Young et al., "The Neural Basis"; L. Young and R. Saxe, "The Neural Basis of Belief Encoding and Integration in Moral Judgment," *NeuroImage* 40 (2008), 1912–20.

response. Participants with a low RTPJ response—and a presumably weaker representation of agents' false beliefs and innocent intentions—assigned more blame to agents causing accidental harm (such as Grace, who accidentally poisoned her coworker). Participants with a high RTPJ response blamed agents less for causing accidental harm.[4] This correlation suggests that individual differences in moral judgment (i.e., the assignment of blame or forgiveness for accidents) are due at least in part to individual differences in specialized neural circuitry for reasoning about other people's beliefs and intentions.

Of course, the conflict between mental-state factors and outcome factors may account for why we sometimes find it so hard to forgive. The brain data suggest that the strength of the mental-state representation—how we reason about an agent's belief or intent—helps to determine whether we offer forgiveness even in the face of serious harm. Notably, the conflict between mental-state factors and outcome factors may be resolved quite differently in individuals with compromised mental-state reasoning—as in autism spectrum disorders, including Asperger's syndrome. Our recent work suggests that individuals with Asperger's are more likely to judge harms caused accidentally due to false beliefs as morally wrong.[5] In the absence of robust mental-

4 L. Young and R. Saxe, "Innocent Intentions: A Correlation Between Forgiveness of Accidental Harm and Neural Activity," *Neuropsychologia* 47 (2009), 2065–72.

5 J. Moran, L. Young, R. Saxe, S. Lee, D. O'Young, P. Mavros, J. Gabrieli (equal contributors), "Impaired Theory of Mind for Moral Judgment in High-Functioning Autism," *PNAS*. Published electronically January 31, 2011. doi: 10.1073/pnas.1011734108.

state representations, moral judgment appears to be based on outcome factors, such as the amount of harm that's done.

CHANGING MORAL MINDS

Given what we know about the brain basis of morality, can we alter moral judgment by altering activity in target brain regions? We did just this in a recent study: we changed people's moral judgments by producing temporary "virtual lesions" in their RTPJs, using a neurophysiological technique known as transcranial magnetic stimulation (TMS). TMS induces an electrical current in the brain, using a magnetic field to penetrate the scalp and skull. In this study, we used brain scans to identify the RTPJ in each participant and a nearby control region not implicated in mental state reasoning. We then conducted two separate experiments in which we examined the effects of TMS on the RTPJ and on the control region of each participant. Experiment 1 consisted of two twenty-five-minute TMS sessions ten days apart, during which a participant received TMS to the RTPJ in one session and TMS to the control region in the other. Immediately after each session, participants read and responded to two dozen moral scenarios much like the one featuring Grace. (This task lasted less than twelve minutes, and the poststimulation effects of TMS, given our parameters, lasted from about half to twice the stimulation time.) In experiment 2, we modified the protocol so that new participants received very short (500-millisecond) bursts of TMS while making the moral judgment for each scenario. This allowed us to investigate the effect of disrupting RTPJ activity precisely at the time of moral judgment, after mental state information had been presented and encoded.[6]

As we hypothesized, we found that TMS to the control region made no difference in either experiment. However, TMS to the RTPJ made a significant difference in both experiments: moral judgments were based less on mental states and therefore more on outcomes. TMS to the RTPJ did not *reverse* moral judgments; attempted harms (harmful intent, neutral outcome) were still judged morally worse than accidents (neutral intent, harmful outcome). Crucially, though, disrupting RTPJ activity led to more lenient judgments of failed attempts to harm, based on the neutral outcome, and harsher judgments of accidents, based on the harmful outcome.

Disrupting the neural processes that enable us to represent harmful intent, independent of harmful outcome, changes our moral judgments. Since moral judgments depend on specific neural substrates for processing information about beliefs and intentions, this aspect of morality can be selectively impaired by disrupting the specific neural processes for mental state reasoning.

THE ROLE OF EMOTIONS

The role of mental states in moral judgment appears to be one that we endorse; we regard the rule "It is morally wrong to intend harm" as rational. Is there a role for emo-

---

6 L. Young et al., "Disruption of the Right Temporo-parietal Junction with TMS Reduces the Role of Beliefs in Moral Judgments," *Proceedings of the National Academy of Sciences* 107 (2010), 6753–8.

tions in how we apply this rule? Recent fMRI and neuro-psychological evidence suggests that moral judgments do depend on emotional responses to certain mental states. Moral judgments of attempted harms correlate with activation in the ventromedial prefrontal cortex (VMPC), a brain region for emotional processing situated behind and between the eyes. The VMPC response correlates with the assignment of blame for harmful intentions in the absence of actual harm (e.g., as in a failed murder attempt). Individuals with a high VMPC response assigned more blame for failed attempts than did individuals with a low VMPC response.[7]

Does damage to the VMPC reduce moral blame for harmful intentions? We examined moral judgments made by patients with damage to the VMPC, patients with damage to brain regions not implicated in emotional processing, and healthy participants with no brain damage (all alike in age, gender, and IQ). VMPC patients judged attempted harm as significantly less wrong than the other participants did—and even less wrong than accidental harm. All nine of the VMPC patients we tested showed this same striking reversal of the normal pattern of moral judgments, judging failed attempts to harm as less wrong than accidents—revealing an extreme "no harm, no foul" mentality.[8] In line with previous work showing deficits in emotional processing of abstract versus concrete information, we found that VMPC patients may be unable to

---

7 Young and Saxe, "Innocent Intentions."

8 L. Young et al., "Damage to Prefrontal Cortex Impairs Judgment of Harmful Intent," *Neuron* 65 (2010), 845–51.

generate a normal emotional response to another person's mental state.

THE GHOST OR THE MACHINE?

All of these studies focus on how the brain computes one important aspect of morality: the mental state of the moral agent. In general, moral neuroscience aims to uncover many different aspects of morality in the brain in their full physical glory. The work I've described here and the work of many others have begun to reveal that morality takes up space in the brain, and a lot of it. After all, morality depends on many cognitive functions, such as the ability to reason about people's intentions as well as about the outcomes of their actions and to generate emotional responses to that information. That the moral mind is rooted in the brain may strike us as scary. Is morality all machine and no ghost? Indeed, what we've learned from moral neuroscience so far suggests that how we behave and how we judge other people's behavior can be understood in neural terms—if not now, then eventually. Morality itself, though, may or may not be "all in our heads." Do moral truths, or moral facts of the matter, exist independently of how we think about them? Whether or not they do, the aim of science is to figure out how, in our minds and brains, we attempt to track them down.

DANIEL HAUN
studied experimental psychology in Germany, the United States, and England, completing his PhD in 2007 at the Max Planck Institute for Psycholinguistics. He was a postdoctoral researcher at the Max Planck Institute for Evolutionary Anthropology and subsequently accepted a position as lecturer in developmental psychology at the University of Portsmouth. He is currently directing the Research Group for Comparative Cognitive Anthropology, a joint project of the Max Planck Institutes for Psycholinguistics and Evolutionary Anthropology.

HOW ODD I AM!

DANIEL HAUN

I am different from other people—really different. I know, because I have met people who see the world differently from how I see it. And I don't just mean those with different moral values or cultural practices. I mean people who perceive two lines to be equal in length that I clearly perceive as unequal. People who see differences between colors that look exactly alike to me. Still, I'm not crazy. No more so than other people, anyway. I just happen to be from around here. So are most of my readers, and we all tend to believe that our way of understanding and perceiving the world is the standard, the "natural" way. This is a principle the science of psychology was built on: test a few, know them all.

The truth is, we have no notion of human psychology

on a global scale. More than two-thirds of all participants in psychological experiments are U.S. citizens, and all but a few come from Western industrialized countries.[1] As a consequence, almost all knowledge that psychologists have gained since testing began in the late nineteenth century is based on samples from just over 10 percent of the world's population. With the data we do have from other places, we can already conclude that (1) cross-culturally the human mind varies more than we generally assume and (2) you and I are not even in the center of the distribution. So if we want to understand the human mind, we have to ask a new set of questions: How much cross-cultural variability is there? What are the sources of this variability? Why is it there? What is its common foundation?

Cross-cultural studies have demonstrated that humans from different cultural backgrounds might disagree about which of two lines is longer, whether green and blue are the same color, whether a pile of seven coins is smaller than a pile of eight, and whether a twig is to the left or the north of a pebble. It seems that even some of the most basic domains of human thought—spatial orientation, color perception, numbering—vary cross-culturally.

Here is a simple but striking example. I have worked for several years with a seminomadic hunter-gatherer group in northern Namibia called the ≠Akhoe Hai‖om. Many of the ≠Akhoe Hai‖om maintain their ancient cultural practices, such as healing trance dances and hunting magic. More remarkably, instead of thinking about things in the

---

1 J. Henrich, S. Heine, and A. Norenzayan, "The Weirdest People in the World?," *Behavior and Brain Science* 33 (2010), 61–135.

world as occupying space to their right or left or fore or aft, as we do, the ≠Akhoe Hai‖om use a landscape-based system for spatial orientation. That is, they understand their bodies as moving through space in compass directions. Let's say you wanted to teach the "electric slide" line-dance step to some ≠Akhoe Hai‖om dancers. You'd tell them to start by facing south (try pointing south without a compass; the ≠Akhoe Hai‖om can). Then you'd tell them to take two steps west, crossing the eastern leg behind the western one, then repeat the same steps leading with the eastern leg and moving east. After three steps north and some swaying south to north, there's a turn 90 degrees east. Then they are to start over, crossing the northern foot behind the southern foot and stepping south—and so on.

The ≠Akhoe Hai‖om apply this kind of landscape system to objects in space as well: the stick is north of the pebble. Most important, these spatial references are not only characteristic of how they talk about things but also of how they remember moves and memorize where things are in the world. My colleagues and I recently asked ≠Akhoe Hai‖om children to reproduce a short choreography from memory. The catch was that after the training, we rotated them 180 degrees on their own axis for the test. Most of them maintained the same compass directions they had learned in the training, and thus they moved left where we had taught them to move right. In contrast, German children dutifully repeated the learned right-left movements without considering which way they were facing.

Confronted with this kind of difference, we want to know how it comes about. Psychologists have entertained three ideas that might apply here:

1. Populations differ genetically such that infants are more likely to acquire one system of orientation than another.
2. Humans are flexible at birth, and it is children's cultural context that prompts them to adopt one cognitive preference over another.
3. Humans are genetically prepared to solve certain problems in certain ways, but cultural context can sometimes override an innate tendency.

With regard to supposition 1: Humans are genetically much less diverse than any other species of great ape. A single group of wild chimpanzees contains more genetic variability than all members of the human species taken together. Yet in chimpanzees, only thirty-nine behaviors (such as the presence and absence of nut-cracking behavior or the positioning of hands during mutual grooming) have been shown to differ across populations.[2] Clearly, there are fewer chimpanzee populations than human populations, so comparing the total number of variable behaviors might be unfair. However, we would still expect more than thirty-nine behavioral variants within a single human generation in a single cultural setting. So, whereas cross-group behavioral variation in chimpanzees is interesting in that it presents a precursor to culture in a nonhuman species, it hardly matches the variability we see across human populations. Hence, genetic variability seems an unlikely candidate to account for the whole of cross-cultural variability of mind in humans.

---

2 A. Whiten et al., "Cultures in Chimpanzees," *Nature* 399 (1999), 682–5.

So let's, for now, focus on the other two ideas. To decide between them, we need to investigate the baseline—the innate predispositions of the human mind. Researchers have applied various methodologies to approach this question, most often by testing very young infants. But this approach has two serious limitations. If you were to find that three-month-old infants preferred a certain memory strategy over another, you could not be certain the preference was innate, because they would have already been exposed to the world. Moreover, if they failed to display a particular preference, you wouldn't know whether or not it would kick in later, since—despite a common misconception—"innate" does not necessarily mean "present at birth." Children might be innately prepared to acquire a preference over time.

Therefore, my research group has additionally adopted a second approach, aiming to indirectly infer the cognitive characteristics that human beings might have inherited from our extinct evolutionary ancestors by identifying common ground between us and our closest living relatives, the other great apes.

The biggest problem with trying to study our evolutionary ancestors is that they are all dead, so we rely on the logic of an indirect method from evolutionary biology called ancestral-state reconstruction. The idea is this: If a certain cognitive trait exists in all remaining species of a family with common descent, that trait was, with a certain likelihood, also present in the common ancestor. For example, if five dog breeds with common descent all have long floppy ears, it is likely that their common ancestor did as well. We can compare all living members of our own phylogenetic family, the great apes. Today, five species of great apes still exist: orangutans, gorillas, bonobos, chimpanzees, and humans. We now understand a lot about the degree of relatedness of different animal species, including our own relatedness to our close phylogenetic kin. Orangutans are our most distant ape relatives, while chimpanzees and their sister species, the bonobos, are the most closely related to us, sharing more than 98 percent of our genetic code. We are more similar to chimpanzees than chimpanzees are to gorillas. Just as in the dog-breeds example, the argument here is that any common preferences across all nonhuman great apes are likely to have been inherited from their common ancestor and therefore will form a common innate baseline of departure for humans. So in our research group, we have compared performance on identical psychological tasks across all great ape species.

At this point, it is worth remembering the profound differences in spatial memory between people in the Western industrialized world and the ≠Akhoe Hai‖om. While we navigate the world by seeing it in reference to our bodies ("Watch out for that ditch on the right!"), the ≠Akhoe Hai‖om orient themselves based on the landscape surrounding them ("There's a snake just north of your western foot!"). When we investigated spatial memory in all nonhuman great ape species (using methods similar to those used in testing the various human cultural groups), we found that they indeed share a common strategy for remembering where things are. All nonhuman great apes prefer to locate objects in relation to the surrounding landscape rather than to their bodies. According to the logic of ancestral-state reconstruction, this method of using the environment instead of one's body for orientation is most likely the inherited preference in all humans. And whereas some human cultural groups build on and extend this system, other cultural groups, such as those in the Western world, override their inheritance to do something weird. We create a right-left system that even many university students have a hard time using correctly. Everyone knows someone who cannot tell her left from her right reliably. I have two friends who are so bad at telling right from left that when one is doing the driving and the other is in charge of the map, they get where they want to go simply because the driver is misinterpreting the navigator's faulty instructions.

At least where spatial memory is concerned, then, it looks as if all humans inherit a certain preference, which is then either extended or overridden by a process of cultural tuning. Does this mean that Western children start out with an inherited preference for environmental processing of space, which they then have to override? This moves another branch of psychology—developmental psychology—to center stage. Many of the most interesting mechanistic accounts of mental functions were formulated based on observations of their emergence during cognitive development. When you compare child development across different cultures, you find remarkable differences in parenting styles and also in more general sources of enculturation. Observing children can tell you how their cognition and behavior depart from a shared, inherited base and align with that of the people around them—so much so that they find it hard to imagine alternatives. My research group found that German children start out with a preference similar to that displayed by the nonhuman great apes and that this is followed by slow acquisition of a full-fledged right-left solution to tasks.[3] Whereas infant data alone, as noted, allow for only limited interpretation regarding innate predispositions, this combined with studies of older children and with ancestral-state reconstruction of the same tasks lets us interpret our results with new confidence.

By combining cross-cultural studies, cross-species studies, and cross-age studies, we can say something interesting about the extent of cross-cultural variability in human minds and about the inherited behavioral and cognitive foundation common to all humans. But we are still left with an important question: What explains the high degree of population-level variability in human cognition and behavior? This question can be approached on two levels of analysis, proximate and ultimate. At this point in the argument, psychologists may go "Huh?" and biologists may go "Duh!" Let me quickly explain. Proximate explanations account for the causes behind an individual's current behavior—such as, for example, social or environmental influences on the release of a certain hormone. Ultimate explanations are concerned with the *evolution* of behavior. They account for why a behavior came to be— why it was selected throughout our evolutionary history. For example, proximate causes of the blink reflex are a puff of air to the cornea or the electrical stimulation of

---

3 D. B. M. Haun et al., "Cognitive Cladistics and Cultural Override in Hominid Spatial Cognition," *Proceedings of the National Academy of Sciences* 103 (2006), 17568–73.

the trigeminal supraorbital nerve. The ultimate cause of the blink reflex is that protecting your eyes from damage increased your life span and therefore your chances of reproducing. As psychologists often do, I will here deal solely with proximate explanations.

No other animal exhibits anything close to the kind of behavioral variability across populations that the human species exhibits. As mentioned, even the other great apes, who are arguably among the most cognitively flexible non-human animals, display only a relatively small number of stable behavioral differences across groups. This fact becomes even more striking when one considers, again, that humans vary much less genetically than any other great ape species. This gross difference in cross-group diversity is caused by a set of uniquely human characteristics that perfectly set up human children to adapt flexibly to their social/cultural setting.

During human evolution, there were two counteracting trends in the human lineage. While brains kept increasing in size, favoring wide pelvises in females to ensure the infant's safety during birth, bipedalism required increasingly narrow pelvises for stability during locomotion. As a result of these two developments, human children are born at an earlier stage of brain development relative to other great ape species. Thus the infant's head still fits through the mother's narrower pelvis without compromising the trend toward bigger brains. Whereas a newborn chimpanzee's brain is already 45 percent of the average adult size and at one year of age is 85 percent of the adult size, the human brain is only 25 percent of the adult size at birth, increasing to around 55 percent one year later. It takes us up to six years or so to reach 85 percent of the average adult brain size. So the major growth period of the human brain—the time when it is maximally plastic—takes place outside the womb, where children are exposed to their cultural context. Therefore human children may be expected to adopt the local cultural repertoire much more easily than, say, young chimpanzees do.

Furthermore, as human children approach their first birthday, they start frequently engaging in joint activities with caretakers, such as engaging the mother in joint attention to objects or pointing to objects in order to inform. Neither of these two communicative behaviors has been reliably observed in any other great ape species (although this point isn't uncontested).[4] Those activities secure privileged access to cultural information provided by caretakers. Furthermore, social learning seems to proceed differently in humans than it does in the other apes. Though other great apes are successful social learners, they are highly result-oriented—that is, they will aim to reproduce mainly the ends of an action rather than the means. Human children also tend to copy the means of a demonstrated action—how it is performed. It is in the "how" that the culturally significant information (i.e., "This is how we do it") is mainly transmitted. In human children, this tendency to concentrate on the means emerges in the second year of life and peaks between the ages of three and five,

---

4 Michael Tomasello, *Origins of Human Communication* (Cambridge, Mass.: MIT Press, 2008); D. A. Leavens and T. P. Racine, "Joint Attention in Apes and Humans: Are Humans Unique?" *Journal of Consciousness Studies* 16 (2009), 240–67.

by which time they will essentially imitate indiscriminately anything caretakers might demonstrate, even when the demonstrated action is clearly inferior to an obvious alternative as far as its results are concerned. This behavioral tendency leads to a quicker and more precise adoption of a culturally specific repertoire.

Finally, human social behavior within and across groups follows particular patterns, which also promote stable cross-group diversity. Within their group, for example, humans conform. With little or no reflection, we adopt the many functionless and ever-changing fads and fashions of those around us, sometimes even when we know the majority to be mistaken. By reducing within-group diversity, conformist behavior stabilizes between-group boundaries, ensuring continued cross-group diversity. Whereas conformist behavior is cross-culturally common in humans and occurs in children as young as four years old, there are, as yet, no convincing demonstrations of conformist behavior in other great ape species (again, this point is not uncontested).[5] Humans tend to arbitrarily dislike characteristics of other groups (so-called out-groups), both the members and their behavior. Besides the tendency to conform to the majority of their own group, the tendency to reject things about other groups adds to the stability of cross-cultural differences. Human diversity and the special character of human psychology contribute to

---

5 A. Whiten, V. Horner, and F. B. M. de Waal, "Conformity to Cultural Norms of Tool Use in Chimpanzees," *Nature* 437 (2005), 737–40; M. B. Pesendorfer et al., "The Maintenance of Traditions in Marmosets: Individual Habit, Not Social Conformity? A Field Experiment," *PLoS ONE* 4 (2009), e4472.

our inevitable tribalism and are cornerstones of our human nature.

In sum, my argument is that humans' slow brain maturation, their early communicative engagements, their unreflective copying of behaviors, their arbitrary discrimination against out-groups, and their general behavior as social sheep uniquely prepare them to display the intriguing variability we see in our species around the globe. This is not to say that conformism and prejudice are right and good. I do, however, argue that they are a part of being human and that they might help explain some differences between us and other animals. At the same time, I hope that understanding the origin and causes of these characteristics will help us overcome them occasionally.

In a few months, I will return to the ≠Akhoe Hai‖om, and while I am busy line-dancing with their children, they'll no doubt look at me amazed by the sheer extent of my oddity.

JOAN Y. CHIAO
is an assistant professor in the Department of Psychology and the Interdepartmental Neuroscience Program at Northwestern University. She received her BS degree (honors) from Stanford University (2000) and her PhD from Harvard University (2006). Her research centers on investigating how cultural factors influence the basic psychological and neural processes underlying social behavior and emotion processing, with an emphasis on integrating psychology and neuroscience research with public policy and population health issues.

WHERE DOES HUMAN DIVERSITY COME FROM?

JOAN Y. CHIAO

Approximately three out of four people living in Africa and Asia are lactose-intolerant—that is, unable to metabolize the sugar found in dairy products—compared with only one in four Europeans. Ashkenazi Jews have a greater-than-normal prevalence of Tay-Sachs disease, while cystic fibrosis is most common among people from northern Europe. Approximately 33 percent of the people living in sub-Saharan Africa carry the gene for sickle-cell disease, whereas that disorder is rare among Europeans. As many as eight out of ten East Asians are genetically susceptible to emotional disorders such as anxiety and depression, yet only one in two Europeans is at a similar genetic risk—yet despite this genetic vulnerability, the lifetime prevalence of such disorders is only 18 percent for Asian Americans, compared with 37 percent for Caucasian Americans. Nevertheless, minorities within the United States often get sick sooner and with greater severity and die sooner than Caucasian Americans. African Americans have higher death rates than Caucasian Americans for twelve of the fifteen leading causes of death, while Hispanics have higher death rates than Caucasian Americans for diabetes, hypertension, and cirrhosis of the liver. Population health disparities are not only puzzling but costly. In 2009, the U.S. Joint Center for Political and Economic Studies reported that the combined cost of population health disparities and premature death between 2003 and 2006 alone reached $1.24 trillion.

Where does human diversity come from? Human diversity has been a rich source of curiosity since the beginning of human history. In his seventh-century *Etymologiae*, the first of the medieval encyclopedias, Isidore of Seville observed that humans are diverse not only in their physical appearance but also in how they think. Centuries later, philosophers such as René Descartes and John Locke renewed the debate on the origins of human diversity in thinking and behavior. During the Age of Enlightenment, the study of human diversity accelerated, with the emergence of two enormously influential but divergent schools of thought: evolutionary biology and modern anthropology.

The conventional theory of evolutionary biology, advanced by Charles Darwin in his *Origin of Species* (1859), posits that organisms adapt to their environment and, over time, through the process of natural selection, attain the traits that best enable them to survive and reproduce in their environment. The concept of natural selection has been enormously influential in the study of human behavior—particularly in evolutionary psychology, which holds that much of human behavior is universal and arises as a by-product of adaptive mechanisms in the mind and brain. By contrast, pioneering anthropologists such as Franz Boas and Margaret Mead favored scientific approaches to culture that emphasized relativism, whereby cultural differences are best understood on their own terms rather than as products of universal biology.

More recently, culture-gene coevolutionary theory describes a complementary process, by which adaptive mechanisms in the human mind and brain evolved to facilitate social group living through both cultural and genetic selection. In particular, the theory posits that cultural traits are adaptive, evolve, and influence the social and physical environments in which genetic selection operates. A prominent example across species is the culture-gene coevolution of cattle milk-protein genes and human lactase genes: the cultural propensity of humans to drink milk has led to genetic selection for milk-protein genes in cattle and for the human gene that encodes lactase, allowing us to digest it. This particular coevolution was especially prominent in Western Europe, explaining in part why only one in four Europeans are lactose-intolerant today.

In my laboratory at Northwestern University, I study how culture-gene coevolution has shaped the human mind, brain, and behavior. My research is part of a growing interdisciplinary field called cultural neuroscience, which combines theory and methods from anthropology, cultural psychology, neuroscience, and population genetics to understand the nature and origin of human diversity. Research in cultural neuroscience is motivated by two intriguing questions about human nature: How does culture (that is, values, beliefs, practices) shape neurobiology and behavior, and how do neurobiological mechanisms (that is, genetic and neural processes) facilitate the emergence and transmission of culture?

The study of culture-gene coevolutionary theory with regard to human behavior has received surprisingly little empirical attention within contemporary fields of psychology and neuroscience. With Kate Blizinsky, a graduate student in my lab, I recently tested whether or not such cultural values as individualism and collectivism, referring to how people define themselves and their relation to others, coevolved with human genes. Individualistic cultures, such as the United States and Western Europe, encourage people to think of themselves as fundamentally independent of one another. By contrast, collectivistic cultures, such as Japan and Korea, endorse the idea of people as highly interconnected. Individualistic cultures emphasize self-expression and the pursuit of individual well-being over group goals, whereas collectivistic cultures favor maintenance of social harmony over the assertion of individuality. Self-construal style affects a wide range of human behaviors, including how we feel, think, perceive, and reason about our environment and the other people in it. By the age of three, children in the United States and China begin to think about themselves in either an individualistic or a collectivistic way—a tendency that is shaped and reinforced by their parents and even their peers. These cultural divergences in East Asian and Western philosophical views of the self are thought to have emerged early in human history, as is evident from the writings of Socrates and Lao-tzu. However, until very recently little was known about the origin of such cultural differences.

The human brain and human behavior are influenced not simply by culture but also by specific genes, such as the serotonin-transporter gene, SLC6A4, which regulates levels of the hormone serotonin in the brain. The serotonin-transporter gene contains a polymorphic region known as 5-HTTLPR, comprising either a short (S) allele or a long (L) allele. Individuals carrying the short allele of the serotonin-transporter gene are prey to negative emotions, including heightened anxiety and attentional bias to negative information, and are at high risk for depression when faced with major life stress, such as interpersonal conflict, loss, or threat. Despite its potentially maladaptive consequences, between 70 and 80 percent of individuals in a typical East Asian population carry the S allele, compared with a typical European sample, in which between 40 and 45 percent of individuals are S carriers. Even more surprising, Blizinsky and I found that the lifetime prevalence of anxiety and depression in East Asia is actually lower than in Europe and North America. What could explain this seemingly paradoxical finding?

Geographic variability in historical and contemporary pathogen prevalence has been found to predict geographic variability in individualistic and collectivistic cultural norms. Nations with a greater historical and contemporary prevalence of infectious diseases, such as malaria, typhus, and leprosy, are more likely to endorse collectivistic cultural norms, perhaps because collectivistic norms may make it easier to mount a defense against pathogens.

By emphasizing conformity over individuality, collectivistic cultures are better able to encourage such norms as reduced social and physical contact with outsiders, allowing for successful containment of infectious disease. Given the adaptive value of collectivist norms, we hypothesized that increased pathogen prevalence in East Asian regions may be associated with increased collectivistic values, which arose due to genetic selection of the S allele of the serotonin-transporter gene.

Additionally, by emphasizing social norms that increase social harmony and encourage giving social support to others, collectivistic cultural values likely serve an "antipsychopathology" function, creating an ecological niche that lowers the incidence of chronic life stress, thus protecting genetically susceptible individuals from environmental risk factors that trigger anxiety and depression. This hypothesis complements broader notions that the cultural values of individualism and collectivism are adaptive and by-products of evolution.

To test this hypothesis, Blizinsky and I examined the global association between 5-HTTLPR and the cultural values of individualism versus collectivism across twenty-nine nations. We found that, indeed, collectivistic cultures were significantly more likely to comprise individuals carrying the S allele of 5-HTTLPR. Consistent with our hypothesis, historical pathogen prevalence predicts cultural variability in the degree of individualism versus collectivism, due to genetic selection of the S allele. Remarkably, we also found, across nations, that an increased frequency of S-allele carriers predicts decreased anxiety and mood disorder owing to increased collectivistic cultural values.[1] Our findings illustrate that gene frequency plays a unique role in explaining global variation in the adoption of cultural norms and is fundamental to any comprehensive understanding of culture.

Taken together, these findings support the notion that the cultural values of individualism and collectivism likely evolved in response to geographic variability of infectious disease: Regions with a higher prevalence of infectious pathogens developed more collectivistic cultural norms. The adaptive nature of collectivistic cultural norms in Asia then allowed for genetic selection of the S allele for the serotonin-transporter gene because people genetically at risk for anxiety and depression can nevertheless thrive in a culture that emphasizes social harmony and reduction of interpersonal stress. Moreover, a genetic predisposition to negative emotion may be adaptive in collectivistic cultures, enabling early detection of another person's anger or fear and avoiding actions or interpersonal situations that might induce negative emotional states in others. In this view, the population-health disparities in anxiety and depression that exist today between European and Asian cultures are due to culture-gene coevolution.

A central claim of culture-gene coevolutionary theory is that once cultural traits become adaptive, genetic selection probably causes refinement of the cognitive and neural architecture in the human mind and brain responsible for

---

1 J. Y. Chiao and K. D. Blizinsky, "Culture-Gene Coevolution of Individualism-Collectivism and the Serotonin Transporter Gene (5-HTTLPR)," *Proceedings of the Royal Society B* 277 (2010), 529–37.

the storage and transmission of those cultural capacities. To test this theory, my lab, in collaboration with a team of Japanese scientists led by Dr. Tetsuya Iidaka at Nagoya University, has been studying how the human brain gives rise to the cultural values of individualism and collectivism.

Most recently, we conducted a series of brain-imaging experiments to learn how the human brain enables people to think of themselves in either individualistic or collectivistic fashion. Earlier studies in social neuroscience had shown that a brain region called the medial prefrontal cortex (MPFC) is engaged whenever people think about themselves, as opposed to thinking about other people. These studies were conducted mostly in the United States and Western Europe, so little was known about whether or not cultural values affect neural responses within the MPFC. To address this question, we measured brain activity using functional magnetic resonance imaging (fMRI) in Caucasian Americans and Japanese while they read either individualistic statements about themselves (e.g., "I am funny") or context-based, collectivistic self-descriptions (e.g., "When I'm talking with my mother, I am funny"). We found that the MPFC showed greater activity when our subjects thought about themselves in a manner consistent with the norms of their culture—that is, individualistic subjects tended to respond more to general statements about themselves, whereas the collectivists tended to respond more to context-based statements about themselves.[2]

---

2 J. Y. Chiao et al., "Neural Basis of Individualistic and Collectivistic Views of Self," *Human Brain Mapping* 30 (2009), 2813–20.

Our next question was, How does the human brain facilitate cultural change? Humans have migrated for thousands of years, and immigrants to a new land must undergo rapid cultural change in order to thrive in new cultural contexts. Earlier research in cultural psychology has shown that people can be aware of multiple cultural value systems simultaneously and that one can be primed, or favored over another, simply by exposing people to information consistent with that system. For instance, mere exposure to icons such as the American flag, or advertisements that promote the virtue of the family over the individual, will prompt multicultural individuals to automatically shift their way of thinking toward the primed culture, as a means of adapting to their environment.

To examine the dynamic nature of cultural influences on neural mechanisms of the self, we again used fMRI brain imaging to measure neural activity in bicultural Asian Americans who were primed with either individualistic or collectivistic cultural values. In this study, bicultural participants were randomly divided into an individualistic group and a collectivistic group, and were given two cultural-priming tasks: they read a short essay and then wrote one of their own. Participants in the individualistic group read a story about a Sumerian warrior chosen by a general because of his individual merit and achievements and then wrote a short essay about how they differed from friends and family. Participants in the collectivistic group read a story about a Sumerian warrior chosen by a general because he belonged to the general's family and would bring honor to the family and then wrote a short essay about what their families and friends expected of them.

After this cultural priming, we measured their brain activity while they evaluated either individualistic or collectivistic self-descriptions. Consistent with our predictions, we found that temporarily heightening awareness of individualistic and collectivistic values in bicultural individuals (here, bicultural Asian Americans) modulates neural activity within the MPFC and the posterior cingulate cortex (PCC) in a manner congruent with the cultural priming.[3] These findings are strong evidence that fluctuations in brain response accompany fluctuations in the cultural frame that people adopt—and that such mechanisms help people acculturate to novel social environments.

Our findings, along with others from the growing field of cultural neuroscience, illustrate the vital importance of bridging the social and natural sciences in pursuit of the most elusive questions about our human nature. In a prominent lecture in 1959, the British physicist C. P. Snow condemned the separation of the humanities and the natural sciences into "two cultures," lamenting the division, rather than unity, of knowledge commonplace in universities. Yet it is increasingly apparent that understanding the investigation of complex and pressing phenomena, such as health disparities across populations, requires an interdisciplinary approach that integrates the cultural and biological sciences.

The scientific study of group differences is a dangerous idea. Past attempts to conceptualize group differences as

3 J. Y. Chiao et al., "Dynamic Cultural Influences on Neural Representations of the Self," *Journal of Cognitive Neuroscience* 22 (2010), 1–11.

biological, such as social Darwinism in the late nineteenth century or eugenics in the early twentieth century, fueled group oppression rather than equality, rendering biological explanations for group differences taboo as a scholarly topic. These past lessons teach us that the study of group differences is one of the most challenging for scientists and society alike. To make true progress in understanding human diversity, we must steadfastly resist the human tendency toward ethnocentrism and the demonization of differences.

What's refreshing about the burgeoning field of cultural neuroscience is the potential for us to redress past and contemporary anxieties about human diversity. Through new scientific discoveries, we are continually learning that human diversity is perhaps our most precious ability— humans reinvent themselves time and time again in ways that are adaptive to their immediate cultural environment. Reshaping our notions of human difference will enable us not only to build an interdisciplinary science of human diversity that is truly enlightened but also to direct this knowledge to solving some of the most urgent health-related global problems of our time.

ACKNOWLEDGMENTS

From Vintage Books, I would like to thank Jeff Alexander, my editor, for his continued interest in this series of books, and Danny Yanez; my father and agent, John Brockman, for helping me come up with the idea for this series and convincing me that I could pull it off; and, as with the previous collection, this book would have turned out much differently were it not for Sara Lippincott and her excellent editing.

NICK DEAR

The Villains' Opera

after
The Beggar's Opera
by John Gay

faber and faber

First published in 2000
by Faber and Faber Limited
3 Queen Square London WC1N 3AU

Published in the United States by Faber and Faber Inc.,
an affiliate of Farrar, Straus and Giroux, New York

Typeset by Country Setting, Kingsdown, Kent CT14 8ES
Printed in England by Mackays of Chatham plc, Chatham, Kent

To obtain a copy of the music by Stephen Warbeck,
please apply to Peters, Fraser & Dunlop Ltd., Drury House,
34–43 Russell Street, London WC2B 5HA

A CIP record for this book
is available from the British Library

ISBN 0–571–20509–7

2 4 6 8 10 9 7 5 3 1

The Villains' Opera, with music by Stephen Warbeck, was first presented on the Olivier stage at the Royal National Theatre, London, on 11 April 2000. The cast was as follows:

Mr Big Clive Rowe
Filch Adrian Sarple
Angelo O'Connor Michael Wildman
Warren Liam McKenna
Pinhead David Arneil
Peachum David Burt
Detective Inspector Lockit Oliver Cotton
Detective Constable Mardell Sara Powell
Mrs Peachum Beverley Klein
Polly Peachum Madeleine Worrall
Georgina Allyson Brown
Captain Macheath Alexander Hanson
Steaming Susie Samantha Lavender
Soho Steve Anthony Renshaw
Sledgehammer Perry Jai Armstrong
Catford Jack Anthony Clegg
Laetitia Shiv Pauline Carville
Scotch Gordon James Earl Adair
Harry Paddington Alastair Parker
Ryan Lucas Omar F. Okai
Anne Sally Ann Triplett
Mrs Plumrose Paddy Navin
Dr Plumrose David Alder
Roy Anthony Renshaw
Neil Anthony Clegg
Vixen Leigh McDonald

Freak Neville Robinson
Miss Cinnamon Myra Sands
Chantal Samantha Lavender
Tazelle Ceri Ann Gregory
Girls in Club Helen Latham, Maggie Lloyd-Williams
Desk Sergeant James Earl Adair
Lucy Lockit Elizabeth Renihan
Dora Paddy Navin
Drunk Anthony Clegg
Policemen and Beggars
played by members of the Company

Director Tim Supple
Designer Robert Innes Hopkins
Lighting Designer Paule Constable
Music Director Neil McArthur
Associate Music Director Paul Englishby
Additional Orchestrations Neil McArthur
Director of Movement Jane Gibson
Fight Director Malcolm Ranson
Sound Designer Colin Pink and Simon Whitehorn
Company Voice Work Patsy Rodenburg

THE VILLAINS' OPERA

Act One

SCENE ONE

Mr Big takes the stage. He has a very nice overcoat.

Mr Big
    I am by profession a villain
    And so are my business associates
    I am known to them as Mr Big
    And they to me as pond-life

Good evening, ladies and gents, and welcome to a little musical entertainment we has devised for your pleasure. It won't last all night, you'll be pleased to hear, and there'll be a break in the middle for a couple of swift ones. It'll have a bit of this, a bit of that, and a bit of the other – something of a relief in these uncertain times, know what I mean? So sit back and enjoy. Oh, I should thank the management of this luxurious gaff for allowing us the run of the facilities. It ain't very often that our sort's allowed out of solitary, and to be free to wander through the halls of culture and chat up the proper actresses and that is an absolute thrill. Nice one, Trev. – Just one other thing: should I, during the course of the proceedings, chance to hear any unflattering remarks, there's a very dark car park downstairs, where me and my colleagues will be delighted to discuss your criticisms after the show.

    It ain't our purpose to take the piss
    But to postulate a hypothesis:
    What if there ain't no natural laws
    And the world's just run by teeth and claws?
    What if we're mistaken 'bout civilisation
    And it never happened? My own observation

3

Is this: there ain't one single person
Who for illicit gain shows the slightest aversion –
Under the surface, behind smoke-screens
The whole world's ducking and diving, it seems!

From the plush back seat of a blacked-out Mercedes
London's one long procession of bonkable ladies
From Virginia Water to Tottenham Hale
There are blokes running scams, and the Holy Grail
Is a villa, with pool, some place far distant
Where extradition's non-existent . . .

Our grasp of world affairs is chronic
We look at life through a large vodka-tonic
You tend to miss the political agenda
If you've just woke up from a three-day bender
We ain't depressed, we don't evince
No sign of *schadenfreude* since
Running dope and tarts and embezzling banks
Leaves not much time for millennial angst!

So, gentlemen, ladies, if you ain't averse
To geezers who thieve and slappers who curse
To a tale of a gangster, a fence, a bent copper, a
Corpse or two in our underworld opera
If you ain't particular what you observe
In matters of cheating, and grassing, and luurve
Accompany me down Woolwich, where
A band of villains has their lair . . .

SCENE TWO

*We discover the bar of 'The Flower of Kent'. Filch is
behind the counter, on the phone; otherwise it's empty.
Angelo and Warren, gangsters, enter the pub. They also
have very nice overcoats.*

4

Filch (*into phone*) Right, I'll tell him. No, you stay with your brief, keep your mobile on.

Warren You open?

Filch Just. (*into phone*) Got customers.

Filch puts the phone down. Angelo and Warren go to the bar.

Angelo Where's Macheath?

Filch Eh?

Angelo Mac. Bleeding. Heath.

Filch Don't know him. What'll it be?

Warren Two large vodka-tonics, no ice, slice of lime.

Filch (*setting up the drinks*) Don't think we've got any lime.

Angelo (*sighs*) What a surprise.

Filch Might have a bit of lemon . . .

Warren Haven't you never heard of Tesco's?

Filch Yes.

Warren They sell limes.

Filch Right, I'll tell the gaffer.

Warren Waitrose do and all.

Filch serves the drinks and Warren pays.

Filch Come far, gents?

Angelo Brussels. Dawn flight into City Airport.

Warren Bloody convenient.

Angelo Mind you, taking that Woolwich ferry was a soft idea.

Filch Why?

Angelo 'Cause we've ended up in Woolwich.

Filch Supposed to have a new lease of life, isn't it? With the Dome and that? Urban renewal, we was told.

Warren You was misinformed.

Angelo (*grabs Filch*) Macheath. This is his local, right?

Filch Might be . . .

Angelo Words was said concerning my sister Donnatella. And the words that was said was not nice.

Filch What was they?

Angelo That she does a very good trick with her – no, I ain't repeating it.

Warren Improper advances to his sister. Unforgivable.

Angelo And since me and Warren was passing through horrible Woolwich, we thought we might just say hullo. Being as we know he drinks in this toilet.

Warren Vodka with no bloody lime . . . No wonder they got no trade.

Angelo Give him a message from Angelo.

Angelo waves a flick-knife in Filch's face.

If Macheath ever goes near my sister again, never mind making references to her bleeding anatomy, I'll feed his kidneys to me cat.

Warren Tell him: behave.

Filch is released. Pinhead, a runt in a leather jacket, enters the pub.

Filch He don't come in very often.

Warren Didn't know you got a cat.

Angelo Got two.

Filch Hardly ever really.

Warren Two?

Angelo Black and a white one. I call them Good and Evil. Drink up, let's get back to civilisation.

Warren The boss'll be anxious.

Angelo (*with a grin*) Yeah, but it's sorted, isn't it?

Angelo and Warren knock back their drinks and turn to leave. Pinhead sidles up to them.

Pinhead 'Scuse me, chief. Friend of mine's trying to shift an in-car DVD system – seat-back screens, Dolby, the lot. Unwanted gift. Tempted? Do you a very keen price.

Warren Got one, mate.

Pinhead backs away in the face of Warren's icy stare. Angelo and Warren leave the pub. Pinhead turns to Filch.

Pinhead Bottle of the usual, Filch.

Filch (*pouring*) You don't want to go behind the gaffer's back.

Pinhead Free country, isn't it?

Filch Not round here. You know the score. He can be very particular.

Filch serves a glass of beer. Pinhead scrapes together a few small coins from various pockets and puts them on the bar.

Pinhead If he asks, I've just gone to see a man about a dog. – Keep an eye on that (*his drink*).

Filch Why?

Pinhead Case someone nicks it.

Pinhead exits to the Gents. Filch looks at the empty pub and laughs. Peachum enters from his private quarters. He wears a smart business suit with a buttonhole. He joins Filch behind the bar.

Filch Pinhead's here to see you.

Peachum Pinhead? Is he out? What does he want?

Filch Got some gear, I think. Oh, and Saucy Sherry called. She's in court this p.m., and her brief reckons she's going down.

Peachum In the family way, isn't she? What's the charge?

Filch Soliciting, up Westminster. Fifth offence.

Peachum I'd hate to lose old Sherry. She's led more mugs to a cash machine with her hand on their PIN number than all our other girls put together. Tell her to plead for leniency, on account of a bun in her ancient oven.

Filch Right, Mr Peachum.

Peachum Who's her brief again?

Filch Dead-cert Derry.

Peachum God, no wonder her case is weak, if she's hired that ambulance-chaser. He'd defend Gengis Khan for a crate of Scotch, and everyone knows it. The bench won't believe a word he says.

Filch makes a phone call. Peachum comes out from behind the bar.

THE EMPLOYMENTS OF LIFE

Through all the employments of life
Each geezer abuses his brother;

8

Tart and crook they call husband and wife:
All professions slag off one another.
The cop calls the lawyer a cheat,
The beak says the Met's out of line;
Politicians, because they're so great,
Think their trade is as honest as mine.

Filch returns from the phone.

Filch I give her your orders. She says Tom Gagg was up this morning, and he's been remanded for psychiatric reports.

Peachum He's a waste of space, Tom Gagg. The purpose of a bouncer is to ensure that protection is carried out in a civilised manner, not bash some student's head in. Psychiatric reports! I'll give him psychiatric reports . . .

He takes an electronic organiser from his pocket, opens it on the counter, and proceeds to type into it with his thumbs.

Filch What you writing?

Peachum Memo to his psychiatrist. With a few pertinent details of Tom Gagg's family history. (*He types.*) Thirty years of group therapy for you, mate.

Filch Oh, Angelo O'Connor dropped in. He had a vodka.

Peachum (*exasperated*) Filch, when are you going to learn to prioritise? That is transparently the most important message of the morning. Why the dickens do you leave it till last?

Filch I ain't got a very good memory.

Peachum (*raising his fist*) I'll give you something to remember in a minute. What did he want?

Filch He was looking for Captain Macheath.

9

Peachum Where *is* Macheath?

Filch Ain't seen him for days.

Peachum Angelo O'Connor is a prime-time nutter. He works for Mr Big. What does Macheath think he's doing? He wouldn't be scheming, would he? He wouldn't be hatching a plot?

Filch Dunno, Mr Peachum.

Peachum I'll pull the hairs off his goolies one by one, if he's deceiving me. When he comes in, indicate I want a word.

Filch Right.

Peachum Now, tonight's operation. South-east London has hit the jackpot. Who would've believed it? Greenwich is deluged with unarmed tourists, on account of the giant pancake up the road. The gang have only got to –

Detective Inspector Lockit and his sidekick, Detective Constable Mardell, a black woman, enter the pub.

Inspector. A pleasure to see you. What'll it be?

Lockit While I'm on duty, Mr Peachum? You surely know better than that.

Peachum Eh? Heavy session last night, Lockit?

Lockit (*indicates Mardell*) We never drink on the job, my new assistant and I.

Peachum How do you do, Miss – ?

Mardell Detective Constable Mardell.

Peachum (*shaking hands*) Reginald Peachum. Licensed Victualler and Justice of the Peace. – Filch, get about your business down the cellar.

Filch exits.

Lockit I see the joint is jumping with party animals, as per usual?

Peachum You wait till Happy Hour.

Mardell (*consulting her notebook*) Sir, we have reason to believe that a known felon going by the name of Macheath has been frequenting this establishment. We wish to interview him concerning a number of unsolved incidents in the vicinity.

Peachum Macheath?

Mardell Macheath.

Peachum How long you been posted at Woolwich?

Mardell Two weeks, sir.

Peachum I don't tolerate criminals in 'The Flower of Kent', as the Inspector is perfectly well aware.

Mardell Nevertheless, it is a matter of record that –

Peachum (*to Lockit*) Look, Dick, what's all this in aid of? You can't just barge in here and accuse me of harbouring wide-boys. I run a respectable boozer.

Lockit IT'S A NIGHTMARE
It's a nightmare
They've brung in quotas
I've got to collar
A face or two quick

Peachum Dear oh dear.

Lockit
Down the station
There are wall-charts
Keeping score of
How many we've nicked

Mardell
> There are hints of corruption
> Malpractice and greed
> This Borough disgraces
> The Metropolitan Police

Lockit
> So I've got a new partner
> Who's got a degree
> And the whole of south London
> Is in for a squeeze

Peachum Dear oh dear oh dear.

Mardell
> For the sake of public confidence
> Our image must be pure

Lockit
> For the sake of my pension
> Find me a snout who's keen to grass

Mardell
> Got to go for zero tolerance,
> It is the only cure

Lockit
> Got to make an arrest
> Or get a rocket up the arse!

Peachum Blimey.

Mardell
> It's a scandal
> We don't catch villains
> We've got to show that
> We're on the case!

Lockit
> It's a nightmare
> They've brung in quotas

I've got to send down
A famous face!

So give us a bell if Macheath comes in, will you? He's
top of the pops.

Peachum I shall do my civic duty, Inspector, as I was
ever wont to do.

Mardell We must be ruthless and dedicated. The public
must be persuaded that crime doesn't pay.

Peachum Quite. We don't want anyone getting silly
ideas. Nice meeting you, ma'am.

Peachum takes Lockit aside.

Christ, Dick, who called in Gangbusters?

Lockit The Chief Super! She's got four fucking A-levels,
a BA in something poncey, and for all I know she's
World Cluedo Champion as well. She's trained with New
York's finest, busted Triads in Chinatown, and gone deep
under cover in Thornton Heath.

Peachum Am I meant to be impressed?

Lockit No, but the Chief nearly stained his boxers.

Peachum Get her a transfer to cappuccino-land.
Hampstead or Notting Hill Gate. Stop the poodles
shitting on the pavement.

Lockit No can do. I'm calling in a favour, Reg. Deliver
me up a body by the end of the day – ideally that
bastard Macheath. We have an arrangement, remember?
Wouldn't want to see it come to grief. (*to Mardell, who's
at the door*) Where to now?

Mardell (*consults notebook*) 'The Crooked Billet'.

Lockit sighs, and he and Mardell exit.

Peachum What is the world coming to, when an honest-to-goodness copper can't go on his rounds without having the formbook waved in his face by some ethnic whippersnapper who ten years ago would've been lucky to walk through Woolwich unscathed? That's progress for you, that is. That's the product of one too many Public Inquiries in this neck of the woods. Not that I'm racially whatnot in any way. It's just that we have our own method of doing things round here, and we don't like our centuries-old traditions interfered with by persons who probably support someone else at cricket. I suppose I shall have to offer a sacrifice. Help out a friend in need. And they want the Captain – well, makes hisself out a Captain, on account of derring-do in far-off deserts and the bogs of South Armagh – laughably improbable if you want my opinion. But he's far too important to the gang. I can't do without my second-in-command. Macheath may be a bit fly, indeed he may be up to untold iffiness, but that's my look-out, not the law's. So who's it going to be? Let's have a butchers at the database.

He consults his organiser. We see the members of the Gang as he introduces them.

Catford Jack. Bringing in a very nice line in left-hand-drive executive motors. Change of plates, straight down the Channel Tunnel, and bob's your uncle: sold to a gentleman from Frankfurt. Motoring back the other way is Scotch Gordon, alias Gordon the Cig, alias the Taxpayer's Friend. With a Transit full of rolling baccy. You couldn't have a cheerier companion on the M20 than Gordon, who knows every song ever penned concerning the purple heather and the rumbling sporran. And what he brings into the Treasury is not to be sneezed at. However, given that the Excise lads are wise to his trade, I'd venture his days are numbered. Sledgehammer Perry. There's a question-mark against

Perry. Marble fire surrounds are his forte; been known
to strip a house in Brockley in two hours, between one
lot moving out and the next lot moving in. But. But.
I have heard that he disposes of gear in an irregular
fashion, i.e., not through my good offices. Don't matter
if it's fireplaces or what it is. Rules is rules. Steaming
Susie and Laetitia Shiv. Two very fit lasses from
Plumstead. You should see them go through a crowded
commuter train of a Friday night, blades flashing, boots
flying, and a lovely haul of wallets, mobiles, sparklers
and up-to-the-minute personal hi-fi when they disembark
at the Dockyard. Harry Paddington, a poor old-
fashioned cat burglar, without the least genius. When
will it dawn on Harry that his ability to shin up a
drainpipe, clamber through a lav window, and leg it
down the garden path with a piece-of-shit video from
Curry's is not the kind of crime that is going to gain him
any credit in the twenty-first century? What's more, he
will forget the remote control. Ryan Lucas. Now there is
a problem. Ryan Lucas. Says he wants to give up the
trade and go straight. His dad's got him night-work
cleaning offices, which he calls an honest employment.
Some kids never learn. A spell of juvenile detention
should set him back on the straight and narrow. Soho
Steve, a 'beggar': used to run a lucrative scam
embarrassing TV producers outside the Groucho, till he
developed a taste for Semillon bleeding Chardonnay.
Now he's not only sozzled from morn till night, he's
writing a screenplay as well! Memo: put in a call to
Crimewatch. Got news for you, Steve – you're going to
be on the telly. Got it! Brainwave! I have a mind to give
them . . . Little Tony.

*Unseen by Peachum, Mrs Peachum enters and
immediately fixes herself a large gin, which she downs
in one before immediately fixing another.*

Little Tony has done sweet FA for the last three years. Little Tony has been swanning about Whitehall like the emperor's new clothes. Or in the emperor's new clothes. Whatever the hell the phrase is. It's frigging hard to master, English. Little Tony has been living on his past reputation, at any rate, instead of delivering the goods, and Little Tony has been getting on my tits. He is for the chop.

Mrs Peachum You can't shop Little Tony.

Peachum And why not?

Mrs Peachum He is such a ladies' man. The way he kisses children is a paradigm to the others.

Peachum A what?

Mrs Peachum A paradigm. You're right, you know – your grasp of the language is feeble.

Peachum Please, don't get pissed before lunch. You've got pies to serve.

Mrs Peachum Little Tony's ever so brave, you know, he's done so much for the welfare of the battering classes, and I'd love to run my fingers through his lovely shiny hair. It'll be a terrible shame if he's not allowed to stay in the gang. For his age, he's remarkably virile. Really, Reg, I don't think you should shop anyone. We haven't had a murder for at least six months. And that, my dear, is a blessing.

Peachum What the dickens are you perpetually whining about murder for? If it's got to be done, it's got to be done. It's historical, isn't it? Men's always murdered each other.

Mrs Peachum Well I don't like it. I'm sorry. But I don't.

Mrs Peachum pours herself another drink, and knocks it back.

Peachum It is a thing that raises you up in the eyes of your peers, a murder. Her Majesty's gaols are full of geezers who, on account of having done one, are treated with the utmost respect. So don't go carping about matters you don't understand, all right?

Mrs Peachum All right.

Peachum Jolly good.

Mrs Peachum I don't like it though.

Peachum All right! Have you seen Macheath?

Mrs Peachum Not since yesterday morning. He popped in for some Switch cards he'd left with me – and, you know, he was perfectly charming and agreeable, even though the cards had been stopped. He didn't mind a bit! He's awfully dashing, the Captain. He said one day, if he finishes work in time, he'll come back and watch *Who Wants To Be a Millionaire?* with me and Polly. You know, Reg, I've often wondered – has the Captain got a lot of money stashed away?

Peachum He likes the high life, he does. He won't never get rich. He spends too much time up Mayfair, playing roulette. And that's a nob's game, roulette. If you ain't got an earldom to squander, you might as well forget it.

Mrs Peachum Well, that's a shame, from Polly's point of view.

Peachum What's a shame from Polly's point of view?

Mrs Peachum I think the Captain is soft on the girl.

Peachum Oh yeah? So?

Mrs Peachum Well, if I know anything at all about women, I'd say it's reciprocated.

Peachum What the dickens does that mean, reciperocated? What's she going to do, marry him? He's a small-time blagger and con-man! Blaggers and con-men are generally very good to their tarts, but absolute arseholes to their wives.

Mrs Peachum But what if Polly's in love with him?

Peachum (*aghast*) Love?

Mrs Peachum Yes. And what if he's proposed?

Peachum Look, let me tell you a thing or two, my sweet.

Mrs Peachum Shall we have a drink, if we're talking?

Peachum If you must. (*She must.*) A pretty girl behind the bar is a boon in this line of work. Fellers come from miles around to marvel at her assets. When our Pol makes an appearance, the whole pub plays pocket billiards. Now, I'm entirely happy about this; the lassie enjoys the attention, and I enjoy the profits. But. But. She's got to understand that she's to only lead them on. If she takes a husband, we will be in his power. He will worm his way into our tight-knit family unit. He will learn our secrets!

Mrs Peachum Come, come, my dear, the girl is at an impressionable age, you know. She sees all these models and pop-stars and sportsmen – they all marry young nowadays. It's the fashion. And Polly does like to keep up with the fashion. She'll probably want a big do in church.

Peachum Polly at the altar? With Macheath? And Angelo O'Connor as best fucking man? That is the most disquieting prospect I have ever had unleashed on my imagination. He's making a fair job of ruining my day, and I ain't clapped eyes on him yet! I'm going to get to the bottom of this. You watch the bar. And stop boozing!

Peachum exits into his private quarters. Mrs Peachum pours another drink.

Mrs Peachum . . . I'll just have the one more. He's got a filthy temper, my other half. And he sometimes lashes out. But what makes him think Polly's any different to the rest of her sex, and will only love her husband? And why should Polly's marriage, contrary to all scientific orthodoxy, make her less attractive to other men?

MEN

Men!
All men are thieves in love
Men!
Break in with guns and knives
Men!
Steal other fellers' wives
And women let them smash apart their lives

I love my man
But I fear his hand

Men!
Say sex is all to blame
Men!
Adore a mini-skirt
Men!
Yet beat you if you flirt
And women just get used to getting hurt

I love my man
But I fear his hand

Men!
Will punch you in the mouth
Men!
Send flowers to your door
Men!
Then kick you to the floor
And women keep on coming back for more

I love my man
But I fear his hand

Hope my daughter learns
That love's flame burns . . .

Filch enters from the cellar, lugging a crate of bottles, which he takes behind the bar.

Let me ask you a question, Filch – and don't tell me a fib, for I can't abide fibbing, you know. Is anything going on between Macheath and our Polly?

Filch Oh, don't ask me that, Mrs P – I swore I wouldn't say nothing.

Mrs Peachum So something is going on . . .

Filch She'll have me nuts in the mincer if she finds out. I'm not a grass, that's one thing I'm not.

Mrs Peachum No one's calling you a grass. Just tell me confidentially what knowledge has been earwigged by you in a clandestine manner.

Filch I can't!

Mrs Peachum Filch, I know how you feel about Polly.

Filch Do you?

Mrs Peachum I know you've got her best interests at heart. Unfortunately my husband's going to thrash her black and blue.

Filch Oh, no! That ain't fair.

Mrs Peachum Well, give me the news, and I'll see if I can prevent it.

Filch Will he bash her up if she's got married?

Mrs Peachum No. But he'll certainly bash her up if she's got married and we haven't told him about it.

Filch . . . She was down the Registry Office yesterday.

Mrs Peachum Who with?

Filch Macheath. Might not have got married, of course.

Mrs Peachum No.

Filch But that's where they went.

Mrs Peachum You sure?

Filch I followed them.

Mrs Peachum Did you go in?

Filch No. Got a bag of chips.

Mrs Peachum So where did they go when they came out?

Filch I dunno.

Mrs Peachum You didn't follow them?

Filch Hadn't finished me chips.

Mrs Peachum Filch, I fear you are not destined for great things.

Filch (*sighs*) Yeah, so do I, Mrs P.

Peachum and Polly enter.

Mrs Peachum Reg –

Peachum Polly –

Polly Don't tell me how to use my charms! I know how to use my bleeding charms! I've been causing stiffies since I was in the Brownies, Daddy! It's second nature!

Peachum But Macheath –

Mrs Peachum Reg –

Peachum Shut it.

Polly If I've shown the Captain a few extra courtesies, well, he's certainly made it up to me.

Mrs Peachum Reg, she's –

Peachum I said shut it!

Polly Look at this tattoo! He paid for that! And this bracelet! (*It's round her ankle.*)

Peachum Polly, Polly, my love, if it's in the line of business, I don't care if you're on your feet, on your knees, or on your back! Good gracious no! But if I find out you've done the dirty on me and gone and got married, I'll cut your throat, you slag.

Mrs Peachum Reg, she's gone and got married.

Peachum I thought I told you to – *what*?

Mrs Peachum (*bearing down on Polly*) You inconsiderate slut! I'm so ashamed! How can I hold my head up down the shops? I'll be the laughing stock of Bluewater, I will! (*to Peachum*) She's had a wedding on the quiet. And we weren't even invited!

Peachum I don't suppose the identity of the bridegroom is going to come as a big surprise.

Mrs Peachum Macheath!

Mrs Peachum needs a drink for fortification.

I could have rented a hat! It's years since I rented a hat!

Peachum Christ, he's a cocky little beggar. Altogether too bleeding ambitious.

Mrs Peachum It's the fact she didn't tell me that really hurts. That really hurts, you know! I'm wounded!

Peachum (*to Polly*) Do you think your mother and I would have lived so happily together for so long if we'd been married? Dimwit! Marriage is for morons!

Mrs Peachum She was always an arrogant cow. Now she wants to be a bloody It-girl or something. Didn't have no paparazzi, though, did you? Didn't have no Elton John! I could have got you Elton John, and a white piano, and a medley of his lovely tearjerkers! – But no. Has to slip off on the quiet. And get herself hitched to Macheath.

Peachum He's gone right off-message, he has!

Mrs Peachum God, who'd have daughters? They'd as soon swindle their parents as swindle their social security.

Peachum Cat got your tongue, miss? Are you pukka married, or have you just had the consummation in advance of the service?

> *Polly is silent. She stares at Filch, who keeps his head down. Peachum slaps her. Polly yelps.*

Speak up!

Polly THE THING TO DO
  I thought it was the thing to do
  If you loved him, and he loved you
  I thought it was the thing to do

  I thought it was the proper thing
  You found a bloke, you wore his ring
  I thought it was the proper thing

  I thought it was a lovers' tryst
  When you've courted, and you've kissed
  I thought it was a lovers' tryst

Mrs Peachum
  But Polly, you fool
  You don't have to marry them just 'cause you've
    kissed!

Polly
> I thought it was considered nice
> When you've orgasmed once or twice
> I thought it was considered nice

Mrs Peachum
> But Polly, you fool
> His fingers in your knickers is no reason to get
>   spliced!

Polly
> I thought it was all good clean fun
> What I did, you must have done
> I thought it was all good clean fun!

> What I did, you must have done . . .

Mrs Peachum Yes but never with a gangster! And even if I did I never married them! A shag is a shag, that's all! I think I'm going to need a very large drink!

Peachum has been thinking. He blocks his wife's route to the bar.

Peachum Now let us all please calm down. Having a heart attack ain't going to help anybody. (*He draws Mrs Peachum aside.*) The kid's besotted. This constitutes a crisis. Let the Captain get his feet under the table, and within a month he'll have enough information to put us away for a very long stretch – should our relations, regrettably, turn sour. We have to take steps.

Mrs Peachum A divorce?

Peachum Don't be daft. I do not want a divorcee as a daughter, thank you very much. There's my good standing as a magistrate to consider.

Mrs Peachum What do you want, then?

Peachum I'll settle for her schlepping up to Durham once a month, with a stack of Agatha Christies and a small lump of hash in her mouth. He must be removed from circulation.

Mrs Peachum . . . Is there a reward in it?

Peachum Money's not the be-all and end-all of everything, my dear. There's the matter of keeping the troops in line. Every now and then some bastard has to be made an example of. Leave it to me, I'll sort it. You go indoors and put your feet up – you've had a profound emotional shock.

Peachum approaches Polly. Mrs Peachum watches.

Why so melancholy, Polly? What's done can't be undone – so we'd better all make the best of it.

Peachum gives Polly a hug.

There, I forgive you. Who's Daddy's little girl?

Polly Oh, that makes me so happy! All my tears is at an end!

Mrs Peachum A very odd remark, for one so recently married.

Mrs Peachum exits. Pinhead enters from the Gents. He carries a roll of toilet paper.

Polly And you'll square things with the Captain?

Peachum I will sort it, Pol. Believe me, I will sort it. Just as soon as I can find him . . .

Peachum kisses Polly on the brow and turns to leave. Polly looks apprehensive.

I'll see you directly, Pinhead.

Pinhead (*acknowledging*) Mr Peachum.

Peachum exits.

He's got that glint in his eye. (*He shivers.*) Evil as fuck.

Filch Sorry, Pol. The Gestapo got me. But you watch your step. Your father ain't a man you want to aggravate.

Polly DANGER
    Now I'm scared
    What will they do to him?
    Now I'm fearful
    He's in trouble
    He's so beautiful!
    Wild and beautiful!
    And he's sleeping in my room . . .
    His eyes are closed
    His breathing peaceful
    He doesn't know
    There's danger near . . .

Polly goes to leave. She notices Peachum's electronic organiser lying on the counter. She picks it up.

Think I'll take this. For insurance.

Polly exits to the private quarters. Pinhead seems in some discomfort.

Pinhead You know that new curry place on the High Street? The Raj of Krishnapur?

Filch Yeah?

Pinhead Never go there.

Filch Something disagree with you?

Pinhead I had a chicken jalfrezi that'd unblock the Blackwall Tunnel on a Friday night. Me guts is like Dante's Inferno. (*Hands Filch the toilet roll.*)

Filch What you want me to do with that?

Pinhead Put it in the fridge for us, will you? I'll have it back in half an hour. – Hullo . . .

Georgina, a teenage girl, enters the pub. She has a rucksack and a sleeping-roll.

Filch You sure you're in the right place, miss?

Georgina Yes, I'd like a half of lager, please.

Filch What, in here?

Georgina It looked like it's got some atmosphere. Dark and dingy. I thought I'd try it.

Pinhead What's your name, darling?

Georgina Georgina.

Pinhead Pleased to meet you, Georgina. Let me buy you a drink.

Filch No you don't. (*to Georgina*) We don't sell lager.

Georgina It's a pub, isn't it?

Filch We don't sell lager to civilians. It's sort of private. Members only. I don't really think it'd suit. (*to Pinhead*) You keep your hands off.

Pinhead You're too soft-hearted, Filch.

Filch escorts Georgina to the door.

Filch You want to head up Greenwich. There's all kinds of jazzy bars there.

Georgina I know. I fancied something different.

Filch You got a stud in your belly-button?

Georgina Yes.

Filch (*kindly*) Go to Greenwich. The regulars here think body-piercing is something you do to other people, not to yourself.

Georgina What sort of beer do you serve?

Filch Watney's Red Barrel is favourite. Believe me, it ain't your scene. Come on, I'll show you where the bus stop is.

Filch and Georgina exit.

Pinhead (*sighs*) Oh, Filch, Filch. He won't never go up in the world. To go up in the world you need to be an opportunist. You need to take your chances as and when they offers themselves. For instance, when you finds yourself alone in an empty bar with nobody watching the till . . . that's what an opportunist would regard as an opportunity . . .

Pinhead leans over the counter, looks around, and opens the till. Peachum enters silently and sees him.

Fortune is smiling upon me. My star is in the ascendant.

Peachum creeps up behind Pinhead as he's taking the money from the till.

Peachum Your star is up your fundament, my son.

Peachum kicks him in the arse.

Pinhead Mr Peachum!

Peachum What are you doing exactly?

Pinhead I'm just paying for me next drink. Filch left the bar unattended, see, and –

Peachum You're thieving, you little git! (*He kicks him again.*)

Pinhead Well, that's what I do, ain't it? I'm genetically programmed to nick things! Normally you approve!

Peachum Not when you're nicking from me!

Pinhead . . . Ah. No. See what you mean.

Peachum I should fucking coco! You now owe me a very large favour, Pinhead, which I don't think you would dispute!

Pinhead Well . . .

Peachum (*has an idea*) And I need a man for a spot of surveillance, as it goes. How'd you get released so soon, anyhow? Time off for good behaviour?

Pinhead Said I was entering the priesthood. Worked a treat.

From his pocket Peachum produces a mobile phone.

Peachum Know how to work it? Oh, there's a bar on premium-rate services, so don't try calling phone-sex lines in Ghana.

Pinhead (*nervously*) Love to help, Mr Peachum, but – I ain't allowed out after 7 p.m.

Peachum Why not?

Pinhead Curfew. I'm tagged.

Pinhead lifts his trouser-leg to reveal an electronic tagging device strapped to his ankle.

Peachum Well, that's your lookout. I want to know the whereabouts of someone. So you've just joined my gang. (*He hands Pinhead the phone.*)

Pinhead Please, Mr Peachum . . .

PINHEAD'S LAMENT

I can't get a job
The TV's on the blink
The wife's got angina
Them council flats stink
Me car's off the road
It hurts when I pee
Me mum needs a hip-job

29

I won the lottery –
Lost the ticket, though

Me hairline's receding
Me underwear's stained
I ain't been to Clacton
Except when it's rained
The wife's got lung cancer
She's slitting her wrists
The National Health Service
Doesn't exist
Charlton got knocked out
Of the F.A. Cup
I'm persona non grata
In my local pub
Me kids are dyslexic
And riddled with worm
Me dad's got an ear-ache
I want to go home!

Peachum
When will you learn?
You deal with me
You cross me, I get narked
If I get narked
You're in my debt
You obey my commands, you execute
My desires, my appetites, my dreams

Or you can go straight back to bum hell if you like, for
violation of parole. Sit down, Pinhead.

*Pinhead perches gingerly on a bar stool. Peachum
flashes him a convivial smile from behind the counter.*

What'll it be?

SCENE THREE

Polly's bedroom. There are posters on the walls. In her single bed, naked and asleep, is Macheath. Polly enters. She strokes his hair tenderly.

Polly He looks so different in daylight.

Macheath stirs and wakes. He smiles as he sees her.

Macheath ANGELS
    Polly, angel
    Did you miss me
    While I slept?
    Or did you find
    Another lover
    In the big bad world?

Polly
    The universe is empty
    No other man exists
    I think only of you
    Your face, your skin
    The smell of you
    My darling

Macheath
    My angel

Polly . . . You got any intention of getting up?

Macheath Before lunch? I don't think so, love.

Polly My dad could come in any minute.

Macheath leaps out of bed and starts dressing. There are scars on his body.

How long you reckon you'll survive in this racket?

Macheath Don't you worry about me.

Polly I've been in it since I was little. It's brought quite a few of my boyfriends to an early doom.

Macheath I'm special. Got a lucky charm, I have.

Polly What's that?

Macheath You.

They kiss again.

I'll tell you a secret, Pol. I ain't staying in this line of work for very much longer.

Polly (*excited*) You're going to go straight?

Macheath (*caught off guard*) Straight? Well, yeah, I might become a reformed character. Yes I might. I'm just trying to line up one big deal, and then . . .

Polly We're out!

Macheath ONE BIG DEAL
Just one big deal
Got to talk to a face
Just one big deal
Put the pieces in place

Polly
Just one big deal
And we lick up the cream
Just one big deal
Every alley-cat's dream

Macheath
Just one big deal
Got a scheme, can't go wrong
Just one big deal
Been small-time for too long!

Polly
Just one big deal
And we'll walk hand in hand

By hot tropical seas
With our toes in the sand
Drinking tall daiquiris –

Both
Just one big deal!

Polly When?

Macheath Don't you trouble your pretty head. I know a bloke knows a major operator. I've asked for an introduction.

Polly I hope you're not getting out of your depth.

Macheath I'm up for it, doll, well up for it. My moment has come.

Polly Say you had to go on the run – go underground – for a long time, like – you wouldn't forget me, would you?

Macheath Forget you? Forget my Polly? What kind of words is that?

Polly Forget me and sleep with whoever you come across next. I'm just trying to be realistic.

Macheath Oh, well, I can be realistic, if you like.

OVER THE HILLS AND FAR AWAY

I've been around the block a time or two
I ain't a saint
Know what I mean?
But I hope I never drift away from you

Could a footballer forget his Ferrari? Could a back-bencher forget his brown envelope? I couldn't forget you no matter how many beautiful naked women come my way!

If I was floating weightless on the moon
And held you in my arms
I'd have air to breathe!
I'd never come down

Polly

If I was anchored to the ocean floor
And had you by my side
I'd have air to breathe!
I'd never come up

Macheath

And I would love you all the day

Polly

Every night we'd kiss and play

Macheath

If with me you'd fondly stray

Polly

Over the hills and far away . . .

Macheath After all, I married you, didn't I?

Polly (*smiles*) Yeah.

Macheath And why would I have married you if I didn't love you?

Polly To take over the Gang.

Macheath Who put that idea in your head? That's a load of knicker elastic, that is. I married you because you're the one I want to spend the rest of my life with. I married you because you're you, and you're fucking gorgeous. Now, if there's certain beneficial side-effects to the arrangement, all well and good. No one could say I ain't got aspirations. But don't doubt me motives, I've always had good motives, for everything I done. Polly . . .? What's up?

Polly We've got to separate.

Macheath What are you talking about? What's going on?

Polly We got to split up for a while! I don't trust Daddy further than I could spit. I think you're in danger! I don't want to be alone, but you must get out, now!

Macheath What's the old codger up to?

Polly I don't know, but he's plotting something. I was warned!

Macheath Who by?

Polly Filch.

Macheath (*worried*) The word is out already? Macheath is a marked man?

Polly Don't stay another second! Get in the car and drive!

Macheath finishes dressing very quickly. He turns to Polly.

Macheath I can't leave.

Polly One kiss, and then go, please!

They kiss.

Macheath . . . You are a serious kisser. When I come back, we'll go on honeymoon. Till then no other bird will tempt me, I swear.

Polly I'll be gutted, till I see you again . . .! I won't wear no make-up or nothing! Hang on! I nicked this for you. If he finds out he'll kill me.

Polly gives him the electronic organiser.

Macheath What is it?

Polly It's all the details of the Gang and all his doings. (*She turns it on.*) See? Look up 'Licensed Victuallers' Association, North Kent Branch'.

Macheath pockets it.

Macheath Sweetheart, that is the greatest gift a girl could give a bloke: her father's guilty secrets. I shall treasure it.

Polly Use it to keep out of trouble. Now get out that window, and flee!

Macheath FAREWELL SONG
Like a plane disappearing in the sky
I get smaller and smaller
Till I'm gone

Polly
On the runway, waving with a smile
I watch the last of your lights
Then I break down
And tremble
And cry

Macheath exits through the window.

Call me?

But he's gone.

SCENE FOUR

An underground car park. The Gang lounge around a wrecked car, gearing up for their evening's work: Steaming Susie, Laetitia Shiv, Harry Paddington, Sledgehammer Perry, Scotch Gordon, Ryan Lucas, Soho Steve, Catford Jack. There is some sharpening of blades and polishing of knuckle-dusters, etc.

Susie Where's the Captain?

Steve He'll be along.

Perry What's the score with your brother, Jack? Ain't seen him about.

Jack Denis? Had a most unfortunate accident. He's clocked doing a hundred and ten up Shooter's Hill in a lively little Mazda; the old bill give chase; twenty minutes later Denis wraps it round a bollard on the fringes of Gravesend. I goes to bury him and I finds there ain't much left: the surgeons've had half his organs away. Transplants!

Perry He was in the peak of condition, wasn't he?

Jack He was as clean living as you can get. Never touched more than a cup of tea. His kidneys alone would've set you back a couple of grand.

Laetitia Well, when your time's up, your time's up.

Steve But our time is not up, Laetitia. Our time is now. Nobody lives in the present like we do. We are alive to sensation in a tight-arsed world.

Gordon But why is the general public against us? That's what I canna grasp. Are we any more dishonest than the rest of mankind? No way! We're loyal, and decent, and true. And what we win is ours, rightfully ours, by the ancient laws of conquest.

Laetitia We know our purpose in life.

Jack And we're comfortable with it, too. Who else could say that?

Harry Who else could say they have no fear of death? Or life meaning life?

Perry Who else has such a will to work?

Gordon We ought to get the Queen's Award to Industry!

Laughter.

Harry There ain't one of us wouldn't perish for a friend.

Ryan Nor there ain't one of us'd shaft a mate to get a leg up.

Perry Show me a cabinet minister who could claim as much!

Laughter.

Ryan We are for dividing the spoils of prosperity equal. 'Cause everyone's entitled to some of what's in the kitty, ain't they? Ain't they, Gordon?

Susie What we do is, we trim off the excess fat. All these people, they've got too much! It won't hurt them to tighten their belts. They're all greedy. They don't know when enough is enough. Got to have this, got to have that. They're like jackdaws: they acquire things they ain't got no use for. They're worse robbers than what we are if you want my opinion, 'cause they grab more than they need, and don't share it. I don't see no harm in relieving them of some of the surplus.

Soho Steve has been cutting up eight lines of white powder on a mirror. Now he hands a rolled tenner to Susie.

Steve Here, Suse. Have a blast. – Now, we all know our places?

Perry Cutty Sark.

Ryan Observatory.

Laetitia Oh, I love that new Jubilee Line.

Steve Get some of this up your nose. And may it give you courage!

*Soho Steve passes round the mirror and the Gang snort
their dope and make final checks to their weapons.*

The Gang STREET CRIME
> We're tooled up, fit for glorious battles
> We're keyed up, adrenaline in flood!
> We're covetous of all your goods and chattels
> So don't try no heroics or we'll have your blood!

Gordon and Ryan
> We're stationed by a red light on the New Cross
>   Road
> We wait for some old darling driving on her own
> We sidle up all innocent and chuck a fucking brick
> Through the window, nick her Gucci bag, and laugh
>   till we're sick!

The Gang
> Street crime

Perry and Harry
> We don fluorescent jackets and we stand outside a
>   house
> The occupants has gone to work. Oh dear! did we
>   smell gas?
> A pick-axe and a crowbar gets us swiftly in the door
> We has the fucking boiler out, and half the parquet
>   floor!

The Gang
> Street crime

Susie and Laetitia
> Liverpool Street Station is a lovely place to roam
> Around eleven-thirty when the City's heading home
> Futures traders pissed as farts pile on the Chelmsford
>   train
> We get 'em in the toilets and they don't come out
>   again!

The Gang
  Street crime

Steve and Jack
  You know that brand-new footbridge they're
    constructing on the Thames?
  St Paul's will look delightful, but it's miles from end
    to end
  We won't bother with the patter, all that 'stand and
    deliver'
  We'll just bash you with a metal bar and bung you
    in the river!

The Gang
  We're pumped up, from our livers to our eyeballs
  We're psyched up for intrepid acts of war
  We never read our schoolbooks or our Bibles
  So don't try no heroics or you'll get what for!

*There is a sound behind them, and they spin round,
weapons to hand. It's Filch, and he has Pinhead with
him. Pinhead's terrified.*

Laetitia It's Filch.

Steve You don't want to creep up on us, Filch.

Susie Might lose your good looks.

Filch Mr Peachum sent me.

Perry Who's laughing boy?

Filch Pinhead.

Pinhead All right?

Steve We are, yeah. Are you?

Filch He's just done a stretch for – what was it?

Pinhead Fiddling the Family Allowance.

Steve Straight up? Thought you was a Great Train Robber.

The Gang laugh cruelly. They circle around Pinhead.

Filch The gaffer says, he's to come out with you tonight. Try his hand at the secret arts and mysteries. The gaffer says, trust him.

Two of the Gang grab hold of Pinhead, and a third frisks him thoroughly.

Gordon Trust him?

Jack He's clean. Just got a mobile. (*He gives it back to him.*)

Susie Well, he won't do much damage with *that*.

Steve What's your M.O.? Straightforward cosh-an'-run? Or are you a dip?

Pinhead Mainly I sell things what don't exist. That's what I mainly do.

Laetitia You should work in the City.

Pinhead I know, but I couldn't get the hang of them straps to keep your socks up.

Harry I don't like it. I never seen him before and I don't know any of his relatives.

Steve But if Peachum says –

Filch I wouldn't go against his express wishes. If I was you.

Pinhead PINHEAD'S LAMENT – REPRISE
I'm just a slogger
In need of an in
Me gran's got angina
Me wife's in the bin

Me dog got run over
I lost me car keys
Me hi-fi's buggered
I ain't –

We hear a squeal of tyres and a screech of brakes nearby, as a motor comes to a skidding halt. The Gang look round coolly. Macheath enters, pocketing his keys. He affects a military air.

Macheath Afternoon everyone. All present and correct? No need to salute. I have been with you in spirit; but an unexpected incident detained me. – Who's this joker?

Filch Name's Pinhead. Peachum sent him.

Pinhead All right, guv?

Perry He's supposed to come with us.

Macheath Well, if Mr Peachum recommended him, I can't see a problem with that. Can you, Perry?

Macheath grabs Pinhead by the balls.

So long as he's not a grass. You're not a grass, are you, Pinhead?

Pinhead I swear on my mother's life . . .!

Macheath Not an oath I recognise. Who gives a toss for their mother?

Pinhead I swear by almighty God . . .!

Macheath Nope, I never trust men with beards.

Pinhead I just want to nick a few bob!

Macheath (*letting Pinhead go*) He's a good lad. Straight as a die. (*to Pinhead*) Only kidding, mate. You tag along and try and make your fortune.

Pinhead (*wheezing*) Ta.

Steve We's all set to go on duty. I have the forged tickets for the Millennium Experience. Shall you and me be heading up Bugsby's Way, as per the plan? I had a pint with a security guard, Captain, and he assures me there's ten thousand dickheads milling about in there who've thrown caution to the winds. They think they're on their fucking holidays!

Macheath I would've liked to accompany you on patrol, but –

Steve But what?

Macheath I'm afraid I can't.

Ryan You don't expect us to go out without you?

Macheath Yes.

Harry But you're the guvnor!

Macheath Then follow my orders. I'm staying behind.

Steve . . . We haven't lost our bottle, have we, sir?

A nasty pause as Steve faces up to Macheath. The Gang take a step back. Macheath paces around Steve.

Macheath Does anyone here doubt Macheath's courage?

Steve Not if he comes on the job.

Macheath His willingness to put his life on the line?

Steve So we're off, then, are we?

Macheath I didn't drive down here to have my fucking honour besmirched.

Steve Well, let's get about our business up the Dome.

Macheath gets a leg behind Steve and trips him. Steve falls on his back. As Steve hits the ground, he pulls a gun from his waistband, and points it up at Macheath's face. At the same moment, Macheath draws his own gun, and levels it inches from Steve's. A stand-off.

Laetitia Don't kill him, Captain!

Macheath It's a fair fight.

Steve Well, no, it ain't actually. (*Beat.*) Mine's a replica. It don't even click proper.

Macheath Mine's real, Steve.

Steve Yeah, thought so.

Macheath Do you still have misgivings about my ability to command?

Steve Never suspected it for a moment. We know what a top geezer you are. We're just a bit psyched up, see.

Macheath backs off, and lowers his pistol. Steve gets shakily to his feet, helped by members of the Gang.

Macheath As it should be. You have to be alert out there.

Filch There's plain-clothes all over Greenwich.

Perry That's right, you can't trust no one.

Gordon Some of them look almost human.

Jack Can't go nowhere without fear of a hand on your shoulder.

Laetitia I has many a sleepless night.

Macheath Look, you're my troops and I have faith in you. You are diamonds, every one.

Harry But something's getting on your wick, sir, ain't it?

Susie You're a little bit jumpy, too.

Macheath . . . Peachum.

Steve Is he looking to screw us? I'll shoot him through the head!

Steve raises his gun aggressively. They all look at it.

I'll kick him in the head!

Macheath Hold your horses, Steve. Mr Peachum knows the way of the world, and he's a vital cog in our strategic wheel. He runs a sleek and gainful system, like a big corporation what don't pay tax. But he and I have had a slight difference, and till it's sorted I shall be obliged to go under cover. You know my skills of camouflage. A private dispute between the two of us, though, should not disconcert the majority. Stay onside, 'cause if you split from him you ain't a squad, you're just a bunch of shitwits of no fixed abode.

Jack Don't know what we'd do if we had no Mr Peachum. We need him like a prossie needs a pimp.

Macheath For now, you do, yeah. Things can change, however.

Laetitia Like?

Macheath Can't tell you yet, Laetitia. But plans is afoot.

Ryan We can't be disloyal to Mr Peachum.

Macheath No, nor can't you be disloyal to me, can you, Ryan?

Pinhead What are you thinking of, Ryan?

Macheath Filch. Steve. (*He beckons them aside. Indicating Pinhead*) Is this gobshite kosher?

Filch I dunno. But he'd nick his gran's teeth if he thought he could flog 'em.

Steve Nah, he wouldn't. His gran's got angina.

Macheath Keep an eye on him, Steve. (*to Filch*) What's Peachum's caper? Why is he aggrieved?

Filch I dunno. Perhaps 'cause of what you done to Polly? I dunno.

Macheath Do you know anything, Filch?

Filch Not really.

Macheath So there ain't nothing more you ought to be telling me?

Filch Um . . . not what I can remember, Captain.

Macheath Right. Fuck off. (*to Steve*) I need you to make Peachum think I've left the firm. Which you know I would not do if there was breath left in my body. When the coast is clear I'll send a signal. All right?

Steve Yes, guv.

Macheath (*to the Gang*) Now get out there and rob them blind!

Gordon (*hefting a cudgel*) To war, then!

Macheath sits moodily on the seat of the abandoned car.

The Gang STREET CRIME – REPRISE
    Street crime!
    Street crime!
    Now's the time!
    Street crime!

The Gang exits.

Macheath (*sighs*) Oh, Polly. Why the fuck did I marry her? And I made all them promises, too. If a man loves money, can he be contented with a pound? Well, I love women. Lots an' lots of women.

If there's one sure cure for depression
It's a woman of wild indiscretion.
Her lips, her legs, her hair, her breasts
Enchant me like sweet magic.
I can lose myself in the scent of her,
Amuse myself with contenting her,
Discover the pleasure that she likes best,
And then abscond. It's tragic
But in the grip of this obsession
The game is over with possession –
When she gives herself at last
I move on to greener grass.

I know it's not a thing you're supposed to admit in these enlightened times, but I must have pussy! There is nothing unbends the mind quite like it. The clubs of Mayfair are a magnet for me. On my life, I can't stay away!

He exits. Pinhead emerges from behind the car, where he's been hiding. He dials a number on his mobile, furtively.

Pinhead Mr Peachum . . .?

Steve re-enters, looking for Pinhead.

Steve Oi. You coming?

Pinhead hastily hides his mobile and exits with Steve.

*The garden of a substantial house in rural Kent. Mr Big
enters from his greenhouse, in his gardening clothes,
with a fork.*

Mr Big THE GARDEN OF ENGLAND
>   In the Garden of England
>   I come to rest
>   I till the soil
>   And trouble leaves me

>   I like peat
>   I like mulching
>   I like to touch the roots
>   Of roses in the ground

>   Every man should have a garden
>   A haven from the world
>   A shed, some tools
>   Some fruit trees

>   In the Weald of Kent
>   I live at peace
>   Far from the daily round
>   Of blood and fear

>   *His wife, Anne, a well-preserved lady, enters from the
>   house. She's upset.*

Anne She's gone again! She left a note! (*Reads note.*) 'I
can't stand living here any longer. Don't bother looking
for me. I'm not coming back.'

Mr Big That's the third time this month.

Anne What have we done wrong? Oh, what have we
done wrong?

Mr Big (*shrugs*) She'll sneak home when she's spent her
pocket money.

Anne No, I think she means it. The child is deeply unhappy.

Mr Big So what? You're unhappy. I'm unhappy. Everybody's unhappy! You don't run off and leave a place with five en-suite bathrooms and a snooker room just because you're glum. It's perverse!

Anne You taking that attitude is probably what drove her out. She might have hoped you'd show a little more sympathy, in her hour of teenage crisis.

Mr Big Look, Georgina can have a teenage crisis whenever she wants, so long as she does it indoors.

Anne Don't you care? She's your only daughter!

Mr Big Yeah, I care. Of course I care, Anne.

He takes a mobile from his pocket and hits a button.

Anne What are you doing?

Mr Big I'll put the boys on her tail. They'll soon pick her up.

Anne (*snatches the mobile*) No! How do you think she's going to feel, if a pair of goons sling her in the back of a Merc? She'll be terrified! That's hardly the way to show her we love her!

Mr Big Jesus. I ain't cut out for parenthood.

Anne I've decided. I'm going to find her myself. She'll be heading for Greenwich Market, like she always does. I know where to look.

Mr Big And then what are you proposing to do?

Anne I'm proposing to talk to her. And find out why she hates us so much.

Mr Big She hates me because I said she couldn't have an internet connection in her bedroom. But I'm not going to pamper the girl. You can tell from me. I had to earn my money. And so does she.

Anne Couldn't you be a little more understanding?

Mr Big You know what? This is my afternoon off. And she's fucked it. What I understand is, it's time she sorted herself out, and come to realise what a beautiful home we have provided for her, and what a luxurious fucking lifestyle. I work my fingers to the bone! I criss-cross the city from morning to night, to keep her at that school! I get a few brief hours in my garden, contemplating nature, trying to grasp God's handiwork, and Georgina decides to do yet another runner and spoil it! She's the one who should be a bit more fucking understanding! And you can tell her that from me!

Anne If I ever see her again, I shall.

Anne turns to leave. From the far end of the garden enters Warren, marching Macheath at the point of a gun.

Mr Big Who the fuck is that?

Warren Caught him coming over the fence.

Macheath Good afternoon, sir. Good afternoon, madam. I regret the intrusion, but –

Warren whacks Macheath around the head.

Mr Big I'm gardening! Can't you see?

Macheath I have a proposition –

Mr Big I don't do propositions in the garden! I do roses, and begonias, and privet! Sort him out, Warren.

Anne Not on the property! Go somewhere else.

Mr Big She's right. Somewhere else, Warren.

Macheath It is a beautiful garden. All them flowers, and not a needle or a Durex in sight! Beautiful.

Anne I've no idea how long I shall be.

Anne exits to the house. Mr Big stares after her.

Macheath Sir, I'll only take a moment of your time.

Mr Big Wait a minute – Doctor and Mrs Plumrose are coming for dinner! Anne! I can't entertain on my own!

But she's gone.

Jesus.

Macheath I think this is the nicest garden I've ever set foot in.

Mr Big It ain't bad, is it? I spend a fortune on compost.

Macheath My name is Macheath. I asked Angelo O'Connor to mention me.

Warren Macheath?

Mr Big You know Angelo?

Warren He don't know Angelo from shit. Let me learn him a lesson.

Mr Big Do you know Angelo?

Macheath Well, to say hullo to, yeah. Me and his sister are like that.

Warren Guv, Angelo's been looking for –

Macheath I asked him to tell you I've got a little investment you might be intrigued by. (*He takes the organiser from his pocket.*)

Warren Please, let me break some of his fingers.

Mr Big (*to Warren*) Patience.

Macheath I am addressing Mr Big, aren't I? *The* Mr Big?

Mr Big Round here, I'm a pillar of the community. I'm at the epicentre of village life.

Macheath And who's he? Farmer Giles?

Warren raises his arm again, but Mr Big laughs.

Mr Big Warren, take a breather. I'll speak to this chancer.

Warren retreats crossly to a garden chair, where he lights a fag and reads the 'Racing Post'.

You got one minute, pond-life.

Macheath Mr Big, I have served a long apprenticeship in the industry. I know the rituals and the regulations. I manage a successful little team, up Woolwich, and our year-on-year grosses ain't bad. But my time has now come for elevation. I know in my soul that I was meant for higher things. To put it plain: I want to join your firm.

Mr Big So does every jack-the-lad in London. What you got to offer?

Macheath Two items. Number one, I have in my other pocket – may I reach into my other pocket? – a not inconsiderable wedge of venture capital, what I have been prudently putting aside. (*He produces a thick envelope.*) The name of the game is Class A drugs. Don't tell me I'm wrong 'cause I'm not. I would like to invest in your empire. I would like to establish a fiefdom of my own – answering to your good self, naturally, and my honour is impeccable, ask anybody. Number two – as a gesture of probity and friendship – this little Psion thing contains every detail of every scam going off between the Dome and the Dartford Tunnel. I offer it to you. They're small-fry, yes, they're small-fry – nevertheless a merger with your mighty conglomerate would surely yield a respectable upward blip on the graphs.

Mr Big takes the organiser and starts it up expertly.

'Licensed Victuallers' Association, North Kent Branch.'

Mr Big (*reading*) I see.

Macheath It's yours if you let me in.

Mr Big It's mine anyway.

Macheath I'm asking for a break, sir. My talents are wasted in Woolwich.

Mr Big I see . . .

Warren He could be a plant.

Macheath No, mate. *They* are the plants. *I* am a person.

Warren Ha ha. (*to Mr Big*) He might be under-cover Vice for all we know.

Mr Big A plod wouldn't bung me ten megabytes' worth of little-league muggers and grafters, would he?

Macheath Ain't very likely, is it, Farmer Giles?

Warren He might have turned Queen's Evidence!

Macheath Look, are you looking for some how's-your-father?

Warren Any time, scumbag, any time!

Mr Big Warren, I've asked you politely: back off.

Macheath Yeah, go and dig some turnips.

Mr Big (*to Macheath*) Been inside?

Macheath Four years. Wormwood Scrubs.

Mr Big What's the name of the head screw on G-Wing?

Macheath There ain't no G-Wing in the Scrubs. Only goes up to E.

Mr Big How would you react if I told you to shoot your father-in-law?

Macheath I ain't got a father-in-law. Wait a minute, yes I have. I'd pull the trigger without the slightest compunction.

Mr Big You take drugs? Personally?

Macheath Nope. Not unless you count vodka and tonic.

Mr Big (*grins*) I like your style.

Macheath (*grins*) So does Angelo's sister.

Warren Unforgivable.

Mr Big This device is a right little goldmine. Macheath, I'm going to give you a chance.

Warren Oh, no.

Mr Big Fetch a bottle of champagne, Warren. – Welcome to Kent.

Macheath Ta.

Mr Big offers Macheath his hand, and they shake.
Macheath smiles.

Warren THE LIFE OF CRIME
   I don't get it
   I just don't get it
   I spend sodding hours in the gym
   And up rolls this hood
   Can't beat up a rice pud
   And I'm told to bend over for him!
   Me abs are steel, me pecs are rock!
   The life of crime is all to cock

Macheath Could I have my gun back, Warren, old bean?

Warren hands back Macheath's pistol and then exits
to the house in a profound existential gloom.

Mr Big There's a shipment arriving from the Low Countries. Dawn tomorrow. If you want in, you're in. How much you got?

Macheath Hundred and fifty.

They sit on the garden chairs.

Mr Big (*smiles*) You're hardly going to become the crack king of Catford with a hundred and fifty grand, my son.

Macheath It's a start. I'm going to be a millionaire one day.

Mr Big Millionaire, eh? There's a new definition of a millionaire. Know what it is?

Macheath Nope.

Mr Big It's someone who's *given* a million to charity.

Macheath Have you?

Mr Big A number of good causes benefits from my largesse. And yet, the more I dish it out, the more I rake it in. It's a conjuring trick: money, dope, money, dope, money, money, money. It expands exponentially, like porn-sites on the World Wide Web. Now there's something else we should be thinking about. You any good with modems?

Macheath Hardly, but I know a nice pair of bristols when I see 'em.

Mr Big (*laughs*) Oh, you want to go in at the creative end?

Macheath Wouldn't say no.

Mr Big I value initiative. You pulled a stunt, blagging your way into the Garden of Eden. You could've wound up going over the end of Margate pier. You know that, don't you?

Macheath Yeah.

Mr Big Well make sure you don't fucking forget it. You are my creature now.

Macheath Got you, sir.

Mr Big Good. First thing tomorrow. Light plane, coming down near Woodham Ferrers.

Macheath Where's that?

Mr Big Essex. Warren will give you the details. Be there, on time, with your wedge – I'll set you up in private enterprise. Call it a franchise, shall we?

Macheath I'm a very happy man, Mr Big.

Through the garden enter Dr Plumrose and Mrs Plumrose, a tweedy couple.

Mrs Plumrose Yoo-hoo! Hello there!

Dr Plumrose We too early?

Mrs Plumrose Aren't your hollyhocks divine?

Mr Big Shit. The Plumroses.

Macheath Eh? (*He pulls out his pistol.*)

Mr Big Put it away! They've come for dinner.

The Plumroses THE VILLAGE
 All is well in the village
 The green has been mowed
 There are three ducks on the duckpond
 Four-wheel-drives on the road

 There are scones in the Aga
 The dog's had his walk
 At the lych-gate, by the church-yard
 Old maids meet and talk

Nothing changes in the village
We've found, since Doomsday
That a strict code of convention
Keeps the riff-raff at bay

Warren enters from the house with champagne on a tray.

Dr Plumrose I say, look at that! Champagne!

Mr Big Please, take a seat. A couple more glasses, Warren.

Warren exits to the house.

Dr Plumrose, Mrs Plumrose – this is Mr Macheath, a business associate of mine.

Mrs Plumrose Jolly pleased to meet you.

Dr Plumrose Ditto.

Mrs Plumrose Where's Anne?

Macheath shakes hands nervously. But Mr Big seems quite at ease.

Mr Big Oh, she's just doing a spot of last-minute shopping. She'll be back in time to eat, I'm sure. Indeed, if we're very lucky she'll be back in time to cook it.

They all chortle.

Dr Plumrose (*to Macheath*) So are you in import–export as well?

Macheath Moving into it, yeah. Diversifying, see.

Mrs Plumrose Splendid! Backbone of the nation, trade.

Mr Big We do our bit for the economy, Macheath and I. Now, any sign of that manure you promised me?

Mrs Plumrose I called at the stables this morning. Do you know, the mucking-out girl was listening to the wireless! On headphones! She couldn't hear a word I said.

Mr Big Goodness me. What is rural life coming to?

They laugh again.

Dr Plumrose Damn the manure. I've something more important to raise.

Mr Big Been another outbreak of cheating at the whist drive, Doctor?

Dr Plumrose (*laughing*) No, no, I do hope not!

Mr Big You want me to open the fête again?

Dr Plumrose We'd be delighted. But look here. There's a seat on the Parish Council becoming vacant, due to the sad demise of Colonel Frears, and the chaps and I have been wondering if . . .

Macheath has risen and moved slowly away, holding his champagne glass. The conversation continues, very civilly, with polite laughter, behind him.

Macheath MACHEATH'S AMBITIONS
    Bloody hell!
    I fancy this!
    It's like another planet!
    I want a slice of what he's got –
    The house, the garden, the whole lot!

    Bloody hell!
    I fancy this!
    Neighbourhood Watch and all!
    A green and pleasant land survives
    On the other side of the M25!

    Bloody hell!
    I fancy this!
    Drinkies with the neighbours!
    Sophisticated villainy's
    A model for society!

SCENE SIX

A nightclub in the West End, of the expense-account sort, with banquettes and lap-dancers. Businessmen and gangsters are the principal clientèle.

To one side, two middle managers sit watching. They are Roy and Neil. They were sober once, but that was a while ago. To the other side sits Angelo O'Connor, with his sidekick, Freak.

There's a floor-show: female Dancers greet us with a spectacular routine. Roy stands up, waving a banknote. Vixen steps forward from the Dancers.

Vixen My name is Vixen. What's your pleasure, gents?

Roy Wey-hey! Get a look at those melons!

Vixen Want to see more?

Neil Roy –

Roy Sure do!

Neil Roy –

Roy What?

Neil You've still got your I.D. on.

Roy Oh, bugger.

Roy hastily slips his photo-I.D. over his head and puts it in his pocket.

Vixen You're a couple of snappy dressers, you two. Where do you work?

Neil The BBC.

Roy We've had a works outing. The whole Accounts Department went to the Dome.

Vixen Like it?

Neil Thought it was a bit Mickey Mouse, actually.

Vixen Bet you gave the secretaries a good seeing-to on the coach.

Roy Had a go!

Neil I didn't.

Roy He hasn't had his leg over for years. His missus has his balls locked in the deep-freeze. I've been to conferences all round the world with Neil, and he's never had any extras on his hotel bill, know what I mean? So show him a good time, will you?

Roy tucks a £20 note into Vixen's garter-belt, and puts a hand out to help her on to the table.

Vixen It's forty for a dance.

Roy (*taking out his wallet*) Oh, right. Do you take credit cards?

Here we freeze the action. Vixen doesn't dance; she sings.

Vixen CREDIT CARDS
It's the same all over town
Men getting up, girls going down
Little boys let out to play
Need company, willing to pay

Chorus
A girl must remember the words that best please
   her –
Mastercard, Amex, Diners' Club, Visa

Vixen
It's the same all over town
Men wearing lust's bright purple crown
And we're kept in gainful employ
Soliciting that drop of joy

Chorus
    A lady offended knows what will appease her –
    Mastercard, Amex, Diners' Club, Visa

Vixen
    It's the same all over town
    Men will cough up for a rub down
    They throw away half what they own
    It is a curse, testosterone

Chorus
    Without them a dancer is not a stripteaser –
    Mastercard, Amex, Diners' Club, Visa

All
    Without them a honey will not let you squeeze her
    Without them you are a sad middle-aged geezer
    Without them your balls will remain in the freezer –
    Mastercard, Amex, Diners' Club, Visa
    Mastercard, Amex, Diners' Club, Visa!

The Dancers are showing us what they can do, and so is Vixen. Macheath is discovered among them. He dances like a man possessed.

Macheath I'm on top of the world!

Macheath steps off the dance-floor with a girl on each arm: Chantal and Tazelle. Heading for a table, they pass Angelo.

Angelo? Angelo O'Connor? How's it going, my friend?

Angelo (*to Freak*) We know this jerk?

Freak Small crew south of the river.

Angelo What you doing up West?

Macheath Having a ball, man. It's my finest hour. How's Donnatella?

Angelo Fine.

Macheath Cracking little number, she is.

Freak Take him outside, shall I?

Angelo Wait. What's your name, tosser?

Macheath (*laughs*) Macheath. Get used to it.

Angelo stands, knocking over his chair. Suddenly a knife is in his hand.

Angelo I've been looking for you.

The music grinds to a halt and the Dancers back away. Vixen leaps off the table. Macheath stands his ground.

Macheath Well, I'm a popular bloke.

Freak Sort him, boss.

Macheath We're on the same team, Angelo.

Angelo No we ain't.

Macheath As of today.

Angelo You insulted the name of O'Connor. I've got to give you a smack.

Angelo lunges, and Macheath leaps away. The girls scream.

Macheath Fuck off, I ain't got a weapon!

Angelo You ain't gonna have a tongue when I'm through.

Angelo lunges again.

Vixen Captain!

From inside one of her high boots, Vixen pulls a small flick-knife, which she throws to Macheath. He catches it and faces Angelo. They circle.

Macheath Can I tell you something?

Angelo What?

Macheath Your tits are nearly as big as your sister's. Amazing, that, ain't it?

Angelo lunges again and Macheath fends him off. It's getting serious.

Come on then.

Macheath goes for Angelo. Miss Cinnamon enters quickly and steps between the fighters. She's a matronly lady, with a grey bun. A Bouncer backs her up.

Cinnamon Stop this! What do you think you're doing? Give me that! And that!

She takes their knives and hands them to the Bouncer.

Macheath Sorry, Miss Cinnamon.

Cinnamon I'm surprised at you both! You've read the rules! 'Differences will not be settled in the club!'

Angelo Sorry, Miss Cinnamon. It got out of hand.

Cinnamon People come here to enjoy themselves!

Angelo But he said things about my sister!

Cinnamon That's no excuse! Donnatella's not exactly a nun!

Macheath She goes on her knees nice.

Angelo Watch it . . .

Cinnamon If the pair of you wish to stay here, you must shake hands and make peace.

Angelo Make peace? With him?

Macheath Macheath is renowned for his diplomacy. I am happy to offer my hand.

Cinnamon You're in a good mood, Captain.

Macheath I'm celebrating.

Tazelle He lost seven grand at roulette!

Cinnamon Angelo?

Angelo Yeah, all right . . .

They shake hands.

Macheath As I was trying to tell you, we're compadres now. I've been given a post in your outfit. And everything's coming up roses.

Freak Don't sound very probable, do it, boss?

Angelo Don't sound very probable at all.

Mr Big enters the club with Warren. They approach Angelo's table. Miss Cinnamon greets them with evident pleasure.

Mr Big Miss Cinnamon.

Cinnamon How wonderful to see you.

Mr Big kisses Miss Cinnamon's hand.

Mr Big You're looking lovelier than ever.

Cinnamon I aim to please. Especially if it's you. Bottle of bubbly?

Mr Big (*to Macheath*) Enjoying yourself, Macheath?

Macheath Very much indeed, thank you, sir.

Angelo and Freak glance at each other with disbelief. Macheath grins triumphantly at Angelo.

Mr Big Nice one. – Angelo. Freak.

Mr Big, Angelo, Freak and Warren sit. Miss Cinnamon takes their order for drinks and dispatches a Hostess. Macheath proceeds to a table with Chantal and Tazelle. Vixen watches from a distance. From her other boot she pulls a tiny mobile, flips it open, and makes a call, all the while watching Macheath.

Macheath Join us, Miss Cinnamon! Let's make it a night to remember!

Cinnamon (*to Mr Big*) Back in a minute.

Cinnamon joins Macheath's party. Mr Big begins an earnest discussion with his Gang as drinks are brought.

What's the occasion, then?

Macheath I am a jammy dog, I am. I have wormed my way into the higher echelons of power. I'm well on me way to a glorious fortune . . .!

Tazelle How are you going to make it?

Macheath Exponentially, my love.

Cinnamon Well, I congratulate you, Captain. You've certainly worked for it.

Macheath As you all slog your hearts out for your rewards, I know. Mr Big is right – we are looking particularly glamorous, my dear Cinnamon. Have we had the crow's-feet round our eyes attended to? (*Cinnamon glows with pleasure.*) I thought so. And there ain't a hint of bruising. And Tazelle, my love, why, you are glowing with health. What are you on at the moment? Organic carrot-juice and that? We're not drinking our own bodily fluids again, are we? Well, whatever it is, your skin is radiant with lovely little hormones. I've got to have a discreet feel of it. Oh . . .! And down here especially. Oh . . .! I'm like a nipper in a sweet-shop. I'm going to have to suck a lolly in a minute. Chantal, my delectable Chantal, with the sexy French name, and the lingerie to match. You can climb my Eiffel Tower any time you like, doll. I'll have a bite of your baguette and all. Chantal of the armour-piercing nipples. I can hardly bear to look, but I'll try and force meself. Oh, Lord, I think I've died and gone to paradise . . . birds of paradise, that's what you are . . . somebody give us a kiss . . .

Vixen has joined them. She leans over and kisses Macheath.

Those lips, if I ain't wrong, belong to Vixen.

Cinnamon There's supposed to be a 'No Touching' rule in here.

Macheath Yeah, well I break the rules, don't I? You all know that!

LET'S GO MAD

It is a man's duty
To love every woman that lets him get near
Confronted with beauty
A bloke can't be blamed for a lecherous leer
Let's have fun
While we're young
For when good looks abandon us, we won't get none

The Women
Macheath, you're so handsome
And funny and clever
You hold us to ransom
We just can't resist!
Macheath, you're so charming
And such a good lover
Your smile is disarming
Your lips must be kissed!

Macheath
      I make it my purpose
      To bed every stunner that crosses my path
      And if there's a surplus
      I'm never averse to a *ménage-à-trois*
      Let's go mad
      It ain't bad
      For when lust has deserted us, life will be sad

The Women
      Macheath, you're so handsome
      And funny and clever
      You hold us to ransom
      We just can't resist!
      Macheath, you're so charming
      And such a good lover
      Your smile is disarming
      Your lips must be kissed!

Macheath . . . So who's the lucky girl who's taking me back to their bedsit tonight?

   He shows a wad of money.

Tell you what, I think I can afford all three. Now I've been promoted.

   He laughs happily. The women purr around him.

Here, no offence, Miss Cinnamon.

Cinnamon None taken. (*Sighs.*) I've had my golden years.

   Cinnamon leaves them and goes to Mr Big's table.

Tazelle Are you suggesting that customers can hire us for sex?

Chantal That's against the rules.

Macheath You know my feelings about rules.

They laugh flirtatiously.

Vixen Reckon you could manage all three, do you?

Macheath Well, I've been in training. Match-hardened, you might say.

Chantal I love a man who can go on all night.

Macheath On my life, Macheath never sleeps.

Vixen Would you be interested in a private show?

Macheath A private show? What's that?

Vixen Tazelle and Chantal and me. We perform for you. In private.

Macheath Blimey.

Tazelle But it ain't cheap, lover.

Macheath I'm the last of the big spenders, I am.

Chantal I love a man who isn't stingy with his dosh.

Macheath What's money for? That's what I ask meself. What's it for? It ain't for putting in a pension scheme, that's one thing I'm sure about!

They laugh.

Vixen We get a lot of fellers in here who treat us like objects. But you're different, aren't you?

Macheath I pride meself I can see inside a woman's soul.

Vixen I think you can . . .

Macheath I know what she wants. I've studied it.

Vixen I bet you could give me what I want . . .

Chantal And me . . .

Tazelle And me . . .

Macheath I feel like a fucking god! Honest, I do!

Tazelle You're not a god. You're a gangster.

Vixen So sexy . . .

Chantal With a great big gun.

Macheath (*laughs*) It ain't that big.

Tazelle Go on, let's have a look at it.

Vixen Bet you won't get it out in public.

Macheath You're dead right I won't.

Chantal Go on, I dare you.

Tazelle I'll put it in my mouth. That'd turn you on, wouldn't it? Give you a taster of our little show . . .

Macheath Oh, God . . .

Vixen I don't think he's got a gun at all.

Macheath Yes I have.

Chantal He's got an imaginary gun.

Macheath No, it's a real one! With real bullets and everything!

Vixen So . . .?

Macheath produces his pistol. Tazelle takes it from him.

Tazelle Ooh . . . look at it . . .

Chantal Isn't it nice and hard . . .?

Vixen Shall I call a mini-cab?

Vixen takes her mobile from her boot and speed-dials a number.

Macheath I had that with me in the slums of Sarajevo. Seen me out of many a scrape, that has. What would

I be without my personal side-arm? An insect, that's what I'd be. Crawling on the face of the earth. But look at my achievements! Just me, and one little gun!

Vixen (*into phone*) We're ready now.

Vixen then leans over Macheath and kisses him hard. The club is suddenly invaded by armed Police. They are accompanied by Lockit and Peachum. The music stops.

Lockit There he is!

Police 1 Get him!

The women spring away from Macheath, taking his gun. He is vulnerable and alone. Mr Big and the other Gangsters sit very still.

Macheath What's this about?

Police 2 Don't fucking move!

Lockit You're nicked.

The Police train guns on Macheath. Lockit approaches him. Peachum hangs back. Roy and Neil are quaking in their shoes.

Roy Give a false name, all right?

Neil All right.

Lockit On your feet, Macheath. You'll accompany me back to SE18.

Macheath Betrayed by a kiss, is that it?

He looks to the women, particularly Vixen.

You lousy rotten bitches!

Lockit Didn't think you could *trust* them, did you, lad? What happy pills are you on?

Lockit takes the pistol from Vixen. Macheath is searched.

Police 1 Inspector.

The Policeman hands Lockit the thick envelope which he's taken from inside Macheath's coat. Lockit takes out fistfuls of £50 notes.

Lockit Well, look at that. He's even brought the evidence.

Macheath What do you mean, evidence?

Lockit I am arresting you for armed robbery.

Macheath Don't talk cobblers!

Lockit At 5 p.m. this afternoon, the South Eastern Building Society in Deptford High Street was held up by a lone gunman, and the amount seized was . . . about this much, I should say.

Peachum laughs.

Macheath Don't think I don't know who's behind this. You're a nasty piece of work, Peachum.

Peachum I'm the winner, son. You are the loser.

Macheath I'll be back!

Peachum Yeah, in about twenty years.

Lockit (*to Police*) Take him to the car.

The Police manhandle Macheath towards the exit. They pass Mr Big's table.

Macheath Mr Lockit . . . wait . . . I've got an alibi!

Lockit An alibi?

Peachum An alibi?

Macheath At 5 p.m. this afternoon I was in the presence of this gentleman!

He indicates Mr Big, who remains impassive.

Lockit Is that so, sir?

Macheath In his garden, down in Kent! Nowhere near bloody Deptford!

Lockit Well?

Mr Big IMPORT–EXPORT
    He knows the score
    I ain't never seen him before
    I never clapped eyes on his guilt-ridden face
    I am a man of impeccable taste

Peachum Alibi indeed!

Mr Big
    My associates here
    Will vouch that I was nowhere near
    This Macheath. I don't fraternise with his sort
    I am in import and export

Mr Big hands Lockit his business card.

Lockit Thank you, sir. Enjoy your evening, gents. Ladies. Sorry for the disturbance.

Macheath is frogmarched out. Peachum and Lockit follow.

Peachum I think we can call that a result, Dick.

Lockit I think I can concur with you, Reg. I can now look my superiors in the eye. And your domestic nightmare is over.

Peachum Jolly good. You didn't see any sign of a little electronic gadget when you was searching him, did you?

Lockit What sort of gadget?

Peachum . . . Nah. Don't matter.

Lockit exits. Peachum turns to Vixen, Chantal and Tazelle.

Check your accounts in the morning. The bank might have paid you a bonus.

Peachum winks and exits.

Roy Phew.

Neil I need the toilet.

Neil exits. Mr Big quietly resumes his discussion. Cinnamon goes to the women.

Cinnamon Take a break, girls.

They all sit.

A pity to lose such a good customer.

The women are no longer interested. They relax and smoke.

Vixen Fuck him. Here, Taz, did you finish your dissertation?

Tazelle Yeah. I'm meeting my supervisor tomorrow.

Chantal And then?

Tazelle Well, if they pass me, I've got me MBA.

Chantal Fantastic!

Vixen Yeah!

Cinnamon You'll be leaving us after that, I suppose.

Tazelle I'll be heading for America, I think.

Vixen Yeah, they like an MBA.

73

Chantal California! Lucky you. My boobs will have gone before I graduate.

Vixen How long does it take, Medical School?

Chantal Seven years. If it wasn't for this, I'd never be able to pay the fees.

Vixen Same with RADA. Still, at least I get an audience here.

Chantal You'll get a big break one day.

Tazelle Yeah! The National Theatre!

Vixen (*sighs*) Yeah. Very likely.

Tazelle What about you, Miss Cinnamon?

Cinnamon Me? I'm just an old Soho tart, with a head for business. I grew up in the clip-joints of yesteryear; things were different then. The way you girls operate is quite beyond my ken. All the punters ogling you, and you're totting up the gross national product of Venezuela, or doing heart transplants in your head!

Vixen Times change, Miss Cinnamon.

Cinnamon They sure do.

DIGNITY AND PRIDE

When I was young
I knew my place
Man was my lord and master
But nowadays
A girl assumes
That she's an equal partner

It isn't like it was before, when
We were always fearful
That if we dared step out of line
We'd get a right good ear-full

74

If I was young
I'd do the same
I wouldn't let men use me
But I'm full-grown
Set in my ways
Expect men to abuse me

Chantal, Tazelle and Vixen

It's a matter of dignity
It's a matter of pride
We might be naked on a table
But you don't get inside
Our minds, our hearts, our secret places
You can't hire them on an hourly basis
You might be staring at our arses
But we've plenty left to hide
It's a matter of dignity
It's a matter of pride

Cinnamon

No, it isn't like it was before
You can't be just a simple whore
Any more

Chantal, Tazelle and Vixen

It's a matter of dignity
It's a matter of pride
When we're flattering a punter
Survival is our guide
We are desirable, delicious
But you should be suspicious
We might be lewd, we might get screwed
But when we whispered sweet nothings in your ear
    and told you what a stallion you are –
Hey, boys –
We lied!
It's a matter of dignity
It's a matter of pride!

SCENE SEVEN

Woolwich Police Station. At the front desk, a uniformed Sergeant is taking details from Anne.

Sergeant . . . Christian name: Georgina. How old is she?

Anne Seventeen. I have a photo.

Anne gives the Sergeant a photo. Pinhead is marched in by a Policeman.

Pinhead All right, leave it out, do me a favour, stroll on!

Policeman One absconder, Sarge.

Pinhead I only went out for a packet of fags!

Sergeant You should have more sense, if you're tagged. Once that beeper's gone off, you're history. Take him down.

Policeman Righto.

Pinhead Arseholes.

The Policeman takes Pinhead down to the cells.

Sergeant Sorry. I will file her as missing, madam. But you must understand, we get a lot of runaways reported. And not too many show up.

Anne LIFE WITH A CROOK
    A man came to live
    In a corner of Kent
    He liked the fresh air
    And the tranquillity
    His love for the countryside
    Was evident
    He lived like an heir
    Of the nobility

A girl from a farm
Fell in love with this man
She liked his good looks
He seemed peaceable
Blinded by passion
She did not understand
That life with a crook
Is unspeakable

They married and then
She discovered the worst
Her man was a brute
With no humility
The pleasant relations
That they'd known at first
Turned to absolute
Open hostility

Along came an infant
A girl who'd represent
All the mistakes
Of this matrimony
She was my flower
My flower of Kent
And I'll do what it takes
To bring her back to me
I don't care what it takes
I'll protect my baby

*Mardell has entered and listened. As Anne turns to
leave the station, she intercepts her.*

Mardell You're looking for a stray?

Anne My daughter.

Mardell You know how overstretched we are?

Anne (*nods*) There are lots of real villains out there.

Mardell We often find kids sleeping rough up at North Greenwich. The heat ducts from the new tube station . . . it passes for comfort. We move them on every now and then.

Anne Thanks. Could you show me on the map?

Mardell Sure. Would you like a cup of tea?

Mardell shows Anne a map at the desk. Macheath is led in by the Police, with Lockit and Peachum following. Macheath recognises Anne.
We lose the front desk, and reveal the cells. Anne, Mardell and the Sergeant are gone.

Lockit Welcome to Woolwich, my old chum.

Macheath Home sweet home.

The Police Officers exit.

Lockit All the usual facilities, of course. But we do like to quiz our guests on their personal preferences. Would you like a blocked khazi, or one in full working order?

Macheath I'm skint.

Lockit Twin accommodation sharing with pervert, or rather be on your own?

Macheath I haven't got a cent!

Lockit As to room service: the stew that's been gobbed in, or the gob-free healthy-balance alternative?

Peachum Just lock the bleeder up.

Lockit Would I be correct in assuming that no bung is forthcoming?

Macheath Take pleasure in your work, don't you? You sadist!

Lockit A SMALL RECOMPENSE
    It's a small recompense
    For the grief I endure
    In the pursuit of bastards
    Like you

    Night and day I aspire
    To a little respect
    But I get none from bastards
    Like you

    On TV it's the same
    We are plodding buffoons
    While they glorify bastards
    Like you

    So I've made it my job
    To pull out all the stops
    And humiliate bastards
    Like you

Righty-ho, then: one shagging-shack, awash with effluent, Delia Smith specials not required. Step this way, sir.

Macheath How about me watch? It's a genuine Rolex.

Lockit (*taking the watch*) Yeah, and I'm Barbara Cartland.

Macheath Me jacket?

Lockit (*taking the jacket*) Versace? Fair do's, you can have a single cell. I won't guarantee the lav, mind.

Macheath I'll be out before I need it. You haven't got a shred of evidence.

Lockit Here, Reg – why does evidence always come in shreds?

Peachum Not a clue.

Lockit And what makes an alibi watertight? And why do you get off scot-free?

Peachum I don't like to think about it too much. If I start thinking about it, I find I can barely speak at all.

Lockit It's a riddle, innit?

Lockit opens a cell. He and Macheath enter it.

Peachum Just one thing, before we leave him under lock and key . . .

Lockit There's another. 'Under lock and key.' It's barmy!

Peachum steps up to Macheath and punches him hard. Macheath doubles up. Peachum smashes him on the back of the head. Macheath hits the floor. Lockit runs out to check the coast is clear.

Peachum You have some property of mine, boy!

Macheath I don't!

Peachum You do! My Psion! Where is it?

Peachum starts kicking Macheath. Lockit returns and joins in with gusto.

Lockit (*as he beats Macheath*) What about giving it some welly? Going at it hammer and tongs? And as for knocking seven shades of shit out of you, that's a real puzzler, that is! Mine only comes in one colour!

Mardell enters and is horrified.

Mardell Sir! Stop!

Lockit and Peachum leave Macheath, who crawls into the corner of the cell.

We can't treat suspects like that!

Lockit He ain't a suspect, Mardell, he's the one what did it. Someone answering his description was clearly caught on camera.

He gives Macheath one last dig with his toe-cap.

That was you in the Bart Simpson mask, wasn't it?

Peachum (*to Macheath*) I'll be back.

Lockit and Peachum leave the cell, straightening their ties. Mardell follows them angrily.

Mardell Who do you think you are?

Peachum and Lockit (*together, sonorously*)

WE ARE THE LAW

Justice and might
Guided by God
We have the right
To keep things as they are
The rich rich, the poor poor
As it ever must be
Him and me, we are the law

Mardell Inspector, I would take the liberty of reminding you that a man is innocent until proven guilty, and we are obliged to treat him as such.

Lockit and Peachum stare at her in amazement.

Peachum What planet are you from again?

Lockit Planet Guardian, Reg. Planet Channel fucking Four.

Peachum This is badlands, lady. If we did things by the book, there'd be chaos and anarchy from here to Tunbridge Wells. And if there's one thing they have made it plain they do not want in Tunbridge Wells, it's chaos and anarchy. Understand?

Lockit The rules was made to be bent. Not broken, just bent.

Peachum As any cursory glance at Her Majesty's Government will show you.

Lockit We follows the example of our betters, Mardell. You would do well to do likewise.

Mardell But how are we going to clean up London if we're as vile as the villains?

Peachum Look. You can't change the way things is. It's historical, isn't it? It's the way of the world.

Mardell (*upset*) Then what the bloody hell am I doing here?

Peachum Good question.

Lockit Yup. A real brain-teaser. Definitely one for the panel. (*to Peachum*) Shall we adjourn to my office, imbibe a medicinal brandy, and fine-tune the detail of the fit-up?

Lockit and Peachum exit.

Mardell I'm going out of my mind!

She pulls herself together, and goes into the cell.

I'm sorry about that. Are you OK?

Macheath (*shrugs*) All part of growing up and being British.

Mardell What's that supposed to mean?

Macheath If you grew up where I grew up, you'd know.

Mardell I'm sure I grew up somewhere similar. (*She helps him to sit up.*) It's the stress, I think. You keep it bottled down inside; but sometimes the volcano blows. It's a kind of madness, police work.

Macheath Funny, that – crime's quite the opposite. Quite relaxing, in fact.

Mardell (*smiles*) Plenty of time for peaceful meditation, is there?

Macheath Try it and see. I could use a friend on the force. I've got a lot going off.

Mardell Are you propositioning me?

Macheath Matter of fact I am.

Mardell I see . . .

Macheath You're a good-looking bird, with a nice sense of humour. That's distinctly unusual in a copper. What time you come off shift?

Mardell . . . Now I've heard everything.

Macheath Think about it, all right?

Mardell Yeah.

Mardell leaves the cell, shaking her head in disbelief. She locks the door.

ITALY

Once I went to Italy, and strolled
In wonder through the palaces of gold.
In damp Venetian churches and in Tuscan
Piazzali, at dawn, at noon, at dusk and
Deep into the night, I gazed in awe
On art that seemed divine. I saw
Paradise, glimpses of cosmic beauty,
Emblems of good and bad statecraft, civic duty
Symbolised in every chunk of marble, every swirl
Of paint. But the creators of this world
That looked so pure
Had blood upon their hands

Walked ankle-deep in gore
Assassinated friends
Cut throats and poisoned food
And I felt that I was falling
Down into a formless void
Where reality's appalling
And good comes out of bad
And good comes out of bad
And good comes out of bad

Mardell exits. A pause. It's quiet in the cells.

Macheath
Oh, shit

The steel door opens. Lucy enters with the Sergeant.

Sergeant It's hardly correct procedure, Miss Lockit.

Lucy I know, Sarge, I know. But I could be ever so nice to you when you gets off duty. Couldn't I?

Lucy strokes his face, revealing more flesh than is perhaps necessary.

Sergeant They'll have me on the slab if they find out.

Lucy No one'll find out. Specially not your missus.

Sergeant You're a little tease, you are.

Lucy You loves it, though, doesn't you?

Sergeant Cor . . . Look, I'll give you ten minutes.

Lucy I'll give you a taste of heaven . . .

Sergeant Cor . . .

Lucy . . . if you'll give us the keys.

The Sergeant gives her the keys and exits, closing the steel door behind him. Lucy looks suddenly businesslike.

Right, where is the creep? Where are you, creep?

Macheath Lucy?

Lucy You rotten little snake!

Macheath How'd you get in here?

Lucy How you've got the nerve to look me in the face after what's occurred, I do not know!

Macheath . . . I ain't looking you in the face, Lucy.

Lucy See the consequences of your ungentlemanly acts?

Macheath No, can't say I can. Open the door, there's a love. It don't half pong in here.

Lucy I hope it smells like the bogs of hell, Macheath!

She unlocks the door to the cell. Macheath sits forlornly in his shirtsleeves.

You have robbed me of my quiet. I really hope you suffer.

Macheath Ah, don't say that! Ain't you got a drop of human tenderness, Luce, coming upon your husband in these lamentable circs?

Lucy Husband? Whose husband?

Macheath Why, your husband, my love, in anything but name. In common law, your husband! We can nip down the Registry any time you like, and fill out bits of paper, though I'm not a great one for ceremony, being something of a rebel as you know. But I am a man of honour. And my honour has never been besmirched. By any terms of reference you like, from any society in any of the four corners of the earth, I am, in deed and in practice, your husband.

Lucy Well, isn't that curious. 'Cause I heard you were Polly Peachum's husband!

Macheath Where'd you hear that?

Lucy Never mind!

Macheath No, where?

Lucy Her dad's upstairs with my dad. And I listened at the door.

Macheath Old Peachum and me is having a minor altercation. Ignore everything he says. It's bollocks.

Lucy I should scratch your eyes out!

Macheath Hey – you're not jealous of Polly, now, are you?

Lucy You've married her! I know you have!

Macheath Ooh, pussycat! Miaow!

Lucy And when would I have seen you next, if you had not been busted? Mary Mother of God, how could I fall for a gangster! I feel like I want to scream!

Macheath You ought to go to counselling or something. Why are you so angry?

Lucy You want to know why I'm angry?

Macheath Yeah.

Lucy You want to know why I'm angry?

Macheath Yeah, I said.

Lucy I'll tell you why I'm fucking angry!

Macheath All right.

Lucy I'm three months pregnant, that's why I'm angry!

Macheath That's not a bad reason.

Lucy It's not a bad reason at all!

Macheath Anyone'd be narked.

Lucy Sure it's a heck of a good reason, if you ask me! Well . . .? What have you got to say?

Macheath . . . Any idea who done it?

Lucy You, you fucking little bastard!

Macheath Me?

Lucy You who says he's got an allergy!

Macheath Did I?

Lucy Yes! To rubber johnnies!

Macheath But how can you be sure it's mine, darling? You've been with half of Woolwich.

Lucy Because it's exactly three months since you was last brought in for questioning! You never had no care for me. All what you said was lies. First you gets me up the duff, then you goes and marries Polly Peachum!

THE REAL MCCOY

Down the ante-natal
They said it was a boy
A villain's kiss is fatal
It ain't the real McCoy

It leads to tears and sadness
To trouble and to strife –
While I get morning sickness
You get yourself a wife!

Did I rely on your behaviour?
I must be off my trolley!
When I go into labour
Will you be shagging Polly?

Macheath Lucy, it's a wind-up. Pol, it has to be said, is a bit of a slapper. Now, I have to spend time in 'The Flower of Kent', on matters of business. This you know.

And I have to talk to Polly, and all right I flirt with her on occasion, which is part and parcel of pub etiquette, and indeed expected from a raver like myself. I may even have kissed her to pass the time. I believe I once, accidentally, nuzzled her bum. But she is only putting it about that I married her to wind you up, and bring me down in your estimation, because *she* is jealous of *you*.

Lucy Come off it. I don't believe a word. Polly has put you beyond my reach. You won't never keep your promise.

Macheath Look, the minute I'm out of here, we'll settle it. On the dotted line.

Lucy I ain't falling for that one.

Macheath Do you doubt my honour?

Lucy I doubt everything about you, from your snot to the holes in your socks.

Macheath Well, that's a lovely way to treat a chap who's put away a nice little nest-egg for the day he becomes a dad.

Lucy . . . Have you?

Macheath 'Course I have. It's my one unfulfilled dream. Fatherhood.

Lucy No, you're having me on.

Macheath I want my boy to grow up different from what we did. He'll go to a decent school in Dulwich, and have music lessons, and do his homework, and go skiing with the sons of bankers and doctors and that. I want him to live proper, and not start nicking at the age of three. So I've been saving up, in the fond hope that . . . You've made me very, very happy, Luce. It'd be a privilege to take you as my wife.

Lucy Really?

Macheath Yeah. An honour I don't deserve.

Lucy Correct.

Macheath But you do love me, don't you?

Lucy Yes. For my sins, I do.

Macheath And I love you.

Lucy Really?

Macheath How about a white wedding?

Lucy Really? In church?

Macheath I am definitely ready to give you satisfaction. If you think there is any in marriage. And be a doting dad. What more can I say?

Lucy You ain't married to Polly, then?

Macheath She thinks a very great deal of herself, does Miss Polly Peachum. All them airs and graces, designer-this and label-that – a bit too Fulham Broadway for my taste. A geezer has only to smile at her, and maybe fondle her undercarriage momentarily, and the bimbo reckons she's married. It's pure vanity. There is categorically nothing between us.

Lucy Sure I'd love to get married in church . . .

Macheath Me too. It'd be a first.

Lucy In white. With bridesmaids. And hymns.

Macheath I tell you what, I've found a charming spot down in Kent – weeping willows, oast-houses, golden labradors – the works. I think the church was built by Henry the Ninth. Fancy it?

Lucy Oh yes please!

Macheath That's my girl. (*He kisses her.*) The thing is, I need to get out.

> *The sound of the steel outer doors opening, and Polly enters.*

Polly Where's my poor husband?

Macheath Oh, shit.

Polly Darling! (*She runs to him.*)

Macheath How'd you get in?

Polly The desk sergeant's a pushover. But twenty years, my dove! Daddy says you're going down for twenty bloody years! Oh, my darling! – Here, don't turn away from us. It's your Pol. Remember my bed? My duvet? My teddies? You was only in it this morning!

Macheath Was ever a bloke more hounded?

Lucy Was ever a bloke such a monumental creep!

Polly Oh, my sweet lover, Macheath! Oh, why did I let you go? I won't betray you. I won't divorce you. Each month for twenty years, I'll be there on the dot of visiting time. (*She puts her arms around him.*) What's the matter? Don't push me away! I'm your lawful wedded wife!

Macheath (*to Lucy*) Don't know what's got into her, honest.

Lucy Sure men were born to lie, and women to believe them!

Polly (*to Macheath*) Oi! This is not how I expect to be treated! Behave like a husband should, and kiss me, and say you adore me!

Lucy You fucking little bastard liar!

Polly Oh, the honeymoon is over, is it? Won't you even look at me? Am I so ugly?

Lucy (*to Polly*) Are you truly married to him, then?

Polly Well, from the way he's ignoring me, I'd say it was obvious!

Lucy (*to Macheath*) You was going to have two wives, was you, you monster?

Polly What? Was you marrying her as well?

Lucy And did *she* get a church wedding? In an old-fashioned village in Kent?

Polly Who's been to Kent? Never took me to Kent!

Lucy Never took me anywhere! We did it in the Interview Room!

Macheath Don't get in a two-and-eight, Lucy. I believe she's having a breakdown. You get girls, throbbing with pubescent passion, they can develop this devastating crush on you, and –

Polly Do you have the heart to try and disown me?

Macheath Do you have the heart to try and convince me I'm married, when I know damn well I'm not?

Lucy Really, Miss Peachum, it is unfair to niggle a gentleman when he's up shit creek without any sign of a paddle. You're making an exhibition of yourself.

Polly Common or garden decency, Miss Lockit, ought to have learnt you how to behave with decorum when a husband's holding conversation with his wife. So get your hands off him!

Macheath A joke's a joke, Pol. We all had a laugh. But it's gone too far.

Lucy This is my dad's nick. I can have you slung out in a second.

Polly My dad could have your dad put away for about a million years.

Lucy My dad don't go round fibbing about his daughter's marital state like your dad does!

Polly My dad never told a lie in his life! Your dad perjures himself once a week! And I'm not leaving my husband!

Lucy SLINGING MUD
    Right then, you fat flirt –
    If you want to sling dirt
    If you want to get shitty
    I'll show you no pity!
    You fat flirt!

Polly
    Right then, you fat tart –
    Slinging mud is an art
    There's a knack, and the fact is
    I've had years of practice!
    You fat tart!

Lucy and Polly fight. Macheath struggles to separate them. Polly is deranged with jealousy, and keeps coming back for more.

Macheath Polly! Polly! You're unhinged!

Polly Unhinged? I should think I would be unhinged, the way you've treated me!

Macheath You want some of that Prozac.

Polly I don't want Prozac. (*She bursts into tears.*) I want my mum!

Polly runs out. A pause.

Macheath That girl's got psychiatric problems.

Lucy You did treat her like a man would normally treat his wife.

Macheath But look at it this way: if I was Peachum's son-in-law, why on earth would he have shafted me, and had me slung in here?

Lucy I don't know . . .

Macheath Ludicrous, isn't it?

Lucy . . . Are you trying to tell me, after all that, that you're on the level?

Macheath Yes.

Lucy Sure I don't know what to believe.

Macheath THE GAMBLE
There is no absolute certainty
Who says the earth goes round the sun?
When you're old, you don't want to look back
At the things you might nearly have done
And didn't

It's pointless believing in anything
Religion's a crutch for the weak
When you're old, you don't want to listen
To the words that you wanted to speak
And didn't

Be brave!
Take a risk!
Be bold!

Lucy
When I'm old, will I sit and remember
The good opportunities missed?
Will I count up the days that I frittered away
'Cause I was too rationalist?

93

When I'm old, I don't want to be bitter
When I'm old, I want to rejoice
That I picked up the cup and I threw down the dice
And never regretted my choice!

Macheath
Be brave!
Take a risk!
Be bold!

Lucy
When I'm old, I don't want to be hung up
I want all my grandchildren to see
That I took all my chances with no backward
glances
And nobody manacled me!
I won't be a violet, shrinking
I won't cringe in timidity
When I die they can write this as my epitaph:
She gambled, she won, she was free!

Macheath Are we on, then?

Lucy Swear you love me. Swear.

Macheath I do.

Lucy How happy I'll be, if you're telling the truth!

Macheath I am, Luce. I am. The truth and Macheath are indivisible.

Lucy What do you want me to do?

Macheath Reckon you could get the duty sergeant down here?

Lucy Probably.

Macheath What about your old man?

Lucy He's been boozing with Peachum. After which he generally needs a kip.

Macheath Better and better. Now –

Lucy I'm scared.

Macheath No time for that. Now –

Lucy I'm scared!

Macheath Of what? Getting caught?

Lucy Yes, and scared that you'll run off into the night, and I'll never see you again!

Macheath You know the one reason that I'm coming back? The one reason above all others?

Lucy doesn't. He puts his hand on her belly.

My boy.

Lucy Our boy.

Macheath You've got to get me out, Lucy. Then we can be legally married, and our son can be called Macheath.

Lucy Take me with you.

Macheath In your condition? Don't be thick. But I'll be back, don't worry.

Lucy Then kiss me – kiss me properly – before you go.

They kiss passionately.

Macheath . . . You are a serious kisser.

Lucy So I've been told . . . But that Polly runs in my head strangely.

Macheath Let's not hang about, eh, darling? Do your stuff.

Macheath goes into his cell and pulls the door closed. Lucy reveals even more flesh than is usually thought necessary, and goes to the outer door.

Lucy Sarge? Oh, Sarge!

The Sergeant enters.

Sergeant Finished?

Lucy Not quite. There was a little something I promised you, wasn't there?

Sergeant What, in here?

Lucy No one's around.

Sergeant Cor . . . you're a right little tearaway, you are.

Lucy Sure it's me upbringing. Come closer.

Hesitantly, the Sergeant goes up to Lucy. She unbuckles his belt and drops it on the floor. She starts to undo his trousers.

Sergeant This is the worst thing I've ever done in me entire career.

Lucy You better stand to attention, then, you naughty boy.

Lucy gets on her knees before the Sergeant. Macheath creeps out of his cell. He goes to the Sergeant's belt, takes the nightstick from it, and cracks the Sergeant over the head. The Sergeant hits the deck.

Macheath Keys, quick.

Lucy Proud of me?

Macheath You're a bloody genius.

They make for the outer door. There is the rattle of a tin mug on the bars of an adjoining cell.

Pinhead (*in the cell*) Captain! Captain, it's you, isn't it?

Macheath Who's that?

Pinhead It's me, Pinhead! Take me with you, guv, go on, please! Me wife's got angina!

Macheath Christ.

Pinhead Please, Captain!

Macheath All right! But you're on your own once we're out!

Macheath unlocks Pinhead's cell and releases him.

Pinhead (*to Lucy*) How d'you do.

Macheath Run!

All three run out. Alarm bells ring. Mardell runs in, sees the unconscious Sergeant, and runs out, following Macheath.

SCENE EIGHT

Sirens and alarm bells, and searchlights blazing. Macheath hits the street, still in his shirtsleeves.

Macheath OUT
I'm out!

Pinhead enters behind him and immediately runs off. Gangsters in half-light.

The Gang
On the street, where a villain can feel he's alive
And only the brave and the crazy survive

Macheath
I'm out!

The Gang
In the night, where a man with the drive to excel
Can seek out his rivals, and send them to hell!

Macheath I'm out!

End of Act One.

Act Two

SCENE NINE

Mr Big takes the stage.

Mr Big Do you know what they charge for a vodka-
tonic in here? It's absofuckinglutely diabolical! Warren
nearly had a funny turn. Jesus, just think how much
they'd make if they didn't have to put on plays! Still,
Trev warned us it was pricey, so – mustn't grumble.
We're all having a marvellous time. Aren't we?
(*menacingly*) Aren't we?

A guy in the grip of historical forces
Don't see that his life has no meaningful causes
Or for that matter effects. He sees naught
But the rungs on the ladder, the mythical beanstalk,
The dung at the top of the heap.
He wants to win, and keep
Winning, and be the linch-pin of the nexus
And drive a giant fucking Lexus
And have beautiful maidens to massage his ego
And other parts if so required. To resume: we go
Back now into dark and murky water,
Places gentlefolk like you lot oughter
Steer well clear of after dark. This world is nasty.
The little guy's quick, but no matter how fast he
Ducks and weaves, catastrophe is as a rule
Lurking round the corner. He ain't a moron or a fool,
But he has a disposition to be bossy and
Adopt a motto we might call Panglossian:
Everything turns out for the best.
Well, we'll see. What you reckon? No? Or yes?

SCENE TEN

'The Flower of Kent', later the same night. Filch is behind the bar. The Gang are drinking in a group. They are boisterous.

Harry Whose shout?

Perry Come on, Gordon, you tight bastard!

Jack Get 'em in, Gordon!

Gordon (*to Filch*) Same again all round. Big ones.

Laetitia So I'm pissing up the escalator at Canary Wharf with the geezer's briefcase, right, and he decides he's coming back for afters. Soft git. Built like a jam roly-poly, with his arsey red braces. He tries to vault across. Might have been a whizz broker, but as an athlete, he was pants, I'm telling you!

Jack What happened?

Laetitia He goes shinning down the middle like a nipper on a slide! We wet ourself, didn't we, Suse?

Susie Oh, we laughed. And it's a bloody big escalator, that.

Laetitia Bloody big escalator.

Susie He was a right bell-end.

Steve He hurt hisself?

Susie I fucking hope so.

Steve Ryan?

Ryan Done very nicely, thank you, Steve. I think Mr Peachum will be pleased.

Steve So what about cleaning offices, then?

Ryan Well, for job security and that, this game has it beat hollow, I'd say.

Steve Good man. Gordon?

Gordon Can't complain. Have one yourself, Filch.

Harry And you, Steven? What did you reckon to the Dome?

Steve There's all this interactive stuff, and you can blow a million quid in the Money Zone – it's blinding! And you go inside this human body, and there's mucus and follicles and that, and then these parasites come at you, and –

Harry No, you berk – did you cop a decent haul?

Steve Oh, right. Well, I got a bit carried away, to be honest. If I'd had the Captain with me –

Harry He might've kept your mind on the job.

Laughter.

Susie I hope he's all right, the Captain.

Gordon Aye. He's never let us down.

Jack Had any news, Filch?

Filch shakes his head.

Laetitia What's this master plan he's got?

Filch I dunno.

Perry I tell you one thing: I'll do whatever he asks of me.

Jack Well . . . without Mr Peachum we're dead in the water. You don't want to bite the hand as rocks the cradle, do you?

Steve (*to Filch*) What sort of mood is the gaffer in?

Filch Seems fairly chipper.

The Gang SWEET AS A NUT
Excellent!
Keep him sweet as a nut!
If we bring home the bacon
We might not be taken
Into custody. Which is all right but
You never know, do you?
Never know when your employer will screw you?

*Mrs Peachum and Polly enter from the private
quarters.*

Mrs Peachum What's this – a lock-in? It's quarter past
twelve!

Filch They've had a good outing, Mrs P.

Steve Yeah, look!

*Steve opens a zipped-up grip and shows Mrs Peachum
the haul.*

Mrs Peachum Well, in that case, I'll make an exception.
In fact, I'll break the habit of a lifetime, and join you.
Raise your glasses, everyone. To private enterprise!

The Gang Private enterprise!

They all drink, except Polly.

Filch Chin up, Pol. It ain't the end of the world.

Perry What got into her?

Mrs Peachum CAPTAIN MACHEATH
Captain Macheath
And before, or during, or after
He's stolen her heart away

They all laugh at Polly. Except Filch.

Polly I don't find that very funny.

Mrs Peachum
    Captain Macheath
    He managed to slip in and shaft her
    And the girl is quite distrait

Polly
    I love him!

Mrs Peachum
    Captain Macheath
    Oh Captain Macheath
    As brilliant a lover as he is a thief
    She liked it on top and she liked it beneath
    The wonderful physique of Captain Macheath

All
    Captain Macheath
    Oh Captain Macheath
    As brilliant a lover as he is a thief
    A bull in a field don't compare with his beef
    A definite hard man is Captain Macheath

Polly
    I love him!

Mrs Peachum
    Captain Macheath
    Oh Captain Macheath
    A man who could bring married women to grief
    Were I ten years younger, I'd lift his fig-leaf
    I'd trade in my husband for Captain Macheath

All
    Captain Macheath
    Oh Captain Macheath
    No one complains if he don't use a sheath
    He's had half of London, in bed he's the chief
    We all take our hats off to Captain Macheath!

Polly is isolated and upset. During the song, Peachum has entered.

Peachum Pity he'll be playing with himself for the next twenty years.

Polly Oh! (*She bursts into tears.*)

Mrs Peachum Now look what you've done, you thoughtless man!

Susie Why, what's happened to him?

Peachum I am afraid the Captain has fallen foul of the law.

The Gang glance at each other nervously.

Steve . . . He's nicked?

Peachum Incarcerated good and proper. Committal proceedings in the morning.

Harry Blimey . . .

Peachum So let that be a lesson to you all.

SILLY BUGGERS

There is a line you do not cross
You never undermine the boss
You're just a bunch of third-rate muggers
So don't go playing silly buggers
Have we got that nice and clear?

The Gang
Crystal clear

Peachum
Any other traitors here?

The Gang
No fear!

Peachum Got something for me, then?

Steve hands him the grip.

Jolly good. Look back Friday for your wages. How's the screenplay coming, Steve?

Steve Handsome, ta.

Peachum Good guys win in the end?

Mrs Peachum Whoever wins in the end, you can bet they will be guys.

Peachum Pardon?

Mrs Peachum Well, the girls aren't going to get a look-in, are they?

Peachum What the dickens are you bleating about?

Mrs Peachum Well, it's men fighting turf-wars, isn't it, and women showing their legs. Films are all the same.

Peachum What has that got to do with any single thing we've been discussing?

Mrs Peachum (*pouring another drink*) I know my opinion isn't valued –

Peachum Too bloody right.

Mrs Peachum – but I have a brain, you know.

Peachum Yeah, and luckily it's been pickled in alcohol, so future generations can examine it.

Mrs Peachum Well, no one's going to want a slice of yours, are they? You're hardly going to make the *Who's Who of Modern Criminology*.

Peachum Pardon?

Mrs Peachum (*sighs*) I'll get you a dictionary, Reg.

Peachum (*drawing her aside*) My dear, is something troubling you? Is it the wrong time of the month? Or is the wrong time of the month a thing of the past? Are we in the season of hot flushes?

Mrs Peachum It's Polly. You've wrecked her life.

Peachum You agreed we had to act!

Mrs Peachum Yes, but she's so upset! Poor lamb . . .

Peachum She'll find someone else. I've worked out what the difference is between Polly and an ironing board.

Mrs Peachum Have you really.

Peachum Yes. You can never get the legs open on an ironing board. Our Pol won't go short of rumpy-pumpy. She'll be in love with the next bogroll who takes her to Brighton for the weekend.

Mrs Peachum You do not grasp, Reg, how sincere a woman's ardour can be.

THE RIVER

It's like a river flowing to the sea
Ever slower, ever deeper
Running broader, running sweeter
To an ocean of uncharted mystery

And as the river never fills the sea
Never sates or satisfies it
Never floods it, can't suffice it
So we never quench our thirst for ecstasy

Lockit and Lucy enter. Lucy has a black eye. She's been crying; there's mascara all down her cheeks.

Lockit I have bad news. Macheath has scarpered. And my daughter . . . my daughter . . . aiding and abetting a . . . Holy Christ you're going to get a belting!

Lucy I've already had a belting!

Lockit Well you're going to get another one!

Peachum . . . Are you saying he's escaped?

Lockit He busted out!

Susie Yeah! Oi for the Captain!

The Gang are all pleased. So's Polly.

Polly He's coming for me!

Peachum You shut your gob.

Lucy I think it's me he's coming for.

Lockit How much he offer you, Lucy?

Lucy He promised me happiness, that's all.

Mrs Peachum Hmm . . .

Peachum I am surprised at you, Inspector. You have let the side down in the paternal discipline department. How could you bring up a child to go so against our code of behaviour? Why, it's unnatural!

Lockit I blame her mother, who held a lot of stupid notions about true love this and true love that. She was a right pain in the arse till she divorced me.

Lucy She used to say that love was the only thing in the world worth fighting for.

Lockit Yeah, and she used to get a smack for it, too! Load of claptrap!

Peachum Enough! We have to find Macheath.

Steve He'll be long gone.

Laetitia Vanished into the night.

Ryan He's a right Houdini, the Captain.

Mrs Peachum (*to Lucy and Polly*) Girls, I want a word with you. Come upstairs a minute.

Mrs Peachum, Polly and Lucy exit to the private quarters.

Peachum (*to the Gang*) Right. There has been a deal too much sympathy shown by you lot for my sworn enemy, Macheath. If you know which side your bread's buttered, this will cease forthwith. Or what will happen? Ryan?

Ryan The Inspector will nick us, Mr Peachum.

Peachum Correct. The Inspector will nick you most violently. Now you may eff off. And if you hear a word of him, you call me! Understood?

The Gang troop out. At the door, Laetitia whispers to Susie.

Laetitia But what about the plan?

They exit, muttering to each other.

Peachum They have the loyalty of rattlesnakes. I don't trust them an inch. Give the Inspector a brandy, Filch.

Lockit Thanks. I need it. You know what I just discovered? My daughter's up the spout.

Peachum Blimey.

Lockit Courtesy of our friend.

Peachum You can't count on women, can you?

Lockit No, mate, you can't. First she gets herself pregnant, then she lets him out of gaol!

Peachum She coughed up, did she?

Lockit Bit reluctant at first. But I eventually got a confession.

Peachum He is a rat, isn't he? He marries the one, he impregnates the other. It's as if he don't like us or something.

Lockit What are we going to do?

Peachum Well, we ain't going to panic.

Lockit Taken as read. But what? I want the bleeder!

Peachum I want him and all. He's got me files.

Lockit Your files?

Peachum My personal organiser. Which holds the daily record of our deeds. Contacts. Schedules. Accounts.

Lockit Christ, am I in it?

Peachum How could you not be? You've had enough kickbacks.

Lockit Holy Christ.

Peachum He lifted it, the brazen little fucker. And stashed it who knows where. I had hoped to beat it out of him before he went for trial. But. But. This calls for some revision.

Lockit I am amazed you would commit such things to paper.

Peachum It ain't paper, it's digital. I keep it in me pocket.

Lockit Oh, you do, do you? You wanker! Rule number one – never write nothing down!

Peachum Filch, you can take some refreshments up to the ladies. I reckon they'll be in need.

Filch Ready-mixed cocktails?

Peachum Perfect.

Filch Nuts or crisps?

Peachum Use your discretion.

Filch Bar of chocolate?

Peachum Filch!

Filch hastily piles bottles and snacks on to a tray and exits to the private quarters.

Look here, Mr Lockit. You've done well out of me down the years. Many's the time a little extra's found its way into your pay-packet. And your house is bulging with quality white goods. I've scratched your back generously. And what do I get in return? Rank incompetence, is what! You run that police station worse than a privatised railway!

Lockit Let me tell you something, Peachum. It's only due to my good offices that you're still walking the streets. They're suspicious of you from here to Scotland Yard. I have mislaid evidence, I have altered statements, I have ripped out pages from young plods' notebooks and flushed them down the pan. You owe me your freedom – and your livelihood! Because this boozer would make Marks and Spencer's look like a going concern! So don't start criticising how I run my nick!

Peachum All you had to do was keep him there, till he was sent down on remand!

Lockit Well it wouldn't be such a bleeding disaster, would it, if you hadn't been so thick as to keep a diary!

Peachum You're like an insurance policy what's failed to cough up! I've a mind to let you swing!

Lockit I'll hang your guts out with the washing if you try it! Don't make any plans for a quiet retirement in the sun, Peachum!

A pause as they consider the situation.

. . . 'Mexican stand-off.' There's another one I cannot for the life of me fathom. What's it got to do with blinkin' Mexicans?

Peachum I dunno. – What's it mean anyhow?

Lockit It means we're stood here like a couple of pussies, Reg. With equal force of arms. It means it's a no-win situation.

Peachum Couldn't have put it succincter meself. But while you've been pondering the subtleties of language, I have worked out what we must do.

Lockit What must we do?

Peachum We must kill the runt –

Lockit Perfect.

Peachum – and make it look legit. Can you get your hands on an Armed Response Unit?

Lockit Any squad car. Any squad car is an Armed Response Unit. It's quite hard finding an unarmed Response Unit. There might be a couple at the Chelsea Flower Show.

Peachum Get some sharpshooters. I'll track down Macheath. And we'll set him up. And you'll hit him.

Lockit I can't have my lads fannying around London shooting any villain as takes their fancy. We have standards, you know.

Peachum What about your Lucy? You want to be a grandad to a kid called Macheath?

Lockit . . . We'll kill him.

Peachum and Lockit (*together*)

BOUNTIFUL BULLETS

One shot to the head
And our troubles are over
One hole in his brain
And the business is done
A bountiful bullet
Speeding through space
The work of a moment:
He ain't got a face.
Some say we're cynics
Our ethics is flawed
But remember, remember
We are the law

Peachum Prepare for a stake-out. I'll be in touch.

Lockit (*turning to leave*) Oh, me daughter . . .

Peachum Leave 'em be. They go all funny about killing.

Lockit exits. Peachum pours himself a stiff drink.
Pinhead runs in from the street.

Pinhead Mr Peachum! I've got information!

Peachum Well?

Pinhead It's Macheath! He's escaped!

Peachum (*groans*) Thank you, Pinhead.

Pinhead And I'm up shit creek, sir!

Peachum How frightful.

Pinhead It's me tagging device! It went off! But I executed
a daring escape from Woolwich nick, in order to give
you the news! But now I don't know what to do! If they
catch me, I'm lumbered! And me wife's in terrible health!

Peachum Yeah. (*He has an idea.*) Pinhead, you've been useful to me. I like to look after my own. I have the wherewithal to offer you a brand new identity – someone with a personality, say – and a one-way ticket to the Costa del Sol. Plus a couple of grand thrown in.

Pinhead Would you, Mr Peachum? I'd be made up, I would, if you'd do that for me!

Peachum (*smiles*) There's just one more thing you have to achieve. One last errand, and you can pack your sun-tan oil.

From behind the counter, Peachum produces a small pistol.

Ever handled one of these?

Pinhead (*backing away*) Aw, no. Not that.

Peachum What's the matter? Scared of loud bangs?

Pinhead As it goes, I am, rather.

Peachum And I thought you was hard.

Pinhead VIOLENCE
    I'm a lightweight
    I'm a toe-rag
    I'm an evil little bastard
    I don't play fair
    Don't pay my share
    And simply can't be trusted
    But I ain't a man of violence
    I wasn't made for violence
    I haven't got the bladder for it, see
    There's only one thing for a man like me:
    A life of quiet domesticity

Peachum
    You're a doormat
    You're a scumbag

You're a spotty little urchin
To get away
You have to slay
The man for whom I'm searching
You'll come to terms with violence
It's in our make-up, violence
You'll find you've got a natural tendency
To let off rounds of high velocity
And rid me of my mortal enemy

He hands Pinhead the pistol.

I have a plan to whack Macheath. But should it go
pear shaped – and given the propensity of the boys in
blue to cock things up, we must prepare for such an
eventuality – you're my back-up man.

Pinhead Aw, no.

*Filch enters, with his tray. Pinhead shoves the pistol
into his pocket.*

Peachum Now, the Captain's alone. He'll need his crew.
If he don't set up a meet, I'm a frigging Chinaman.

Filch 'Scuse me. They want refills. They've got all
emotional.

Peachum Bless.

SCENE ELEVEN

*Polly's bedroom. Polly, Mrs Peachum, Lucy. They've all
been crying.*

Mrs Peachum He ain't, he ain't! He's a villain, and
villains don't carry no baggage! He's never coming back.
He's gone!

Polly Not coming back for me?

Lucy Not coming back for me?

Mrs Peachum None of us will ever see him again.

Both the girls start weeping again.

There, there, girls. Just something you've got to get used to. You swim with the sharks, you better expect to get bit.

Polly I've been made such a fool of!

Polly rips up a piece of paper.

Lucy What's that?

Polly Me wedding certificate.

Mrs Peachum Oh, Polly . . .

Lucy There's something I haven't told you. I'm carrying his child.

Polly Oh, Lucy!

Lucy Sure I've been tricked and all. How could I have let myself believe he'd make a decent dad? Must've been dreaming.

Mrs Peachum You poor cow.

Polly I'm really sorry, Luce, that I was so vicious.

Lucy It's all right. How was you to know?

Mrs Peachum What is it about us? Have we got a defective gene or something? Other women meet nice blokes, and live in the suburbs and that. We have to fall for gangsters, and live in a bloody war-zone! It's like we can't resist the stuff that poisons us!

All THE WIVES' COMPLAINT
    We do ourselves no favours
    We let ourselves be gypped
    The men we think our saviours

Are rarely well-equipped
For constancy and virtue
Uxorious they ain't!
They use you and they hurt you!
The manners of a saint
Are not part of their behaviour
They've still got stone-age views
Once dragged back to their cave you're
Just flesh to be abused!

Lucy
I'm carrying his child

Polly
He's got a bloody cheek!

Lucy
I once was meek and mild

Polly
I don't feel mild or meek!

Mrs Peachum
I ask myself the question

Lucy and Polly
We're feeling pretty het up

Mrs Peachum
No answer's to be had

Lucy and Polly
Should not have let him get up

Mrs Peachum
Why do so many women

Lucy and Polly
Our skirts and have a good time

Mrs Peachum
Fall for a bad lad?

Lucy and Polly
It's tantamount to sex crime!

Mrs Peachum
Fall for a wild rover?

Lucy and Polly
We want him to acknowledge

Mrs Peachum
Fall for feckless charm?

Lucy and Polly
He's stirred the bloody porridge!

Mrs Peachum
Fall for a Casanova?

Polly
He's dipped his long thin spoon in!

Mrs Peachum
Fall for a Don Juan?

Lucy
And now I am ballooning!

Polly and Lucy
Christ what a bastard!
Christ what a ponce!

Lucy
A total disaster!

Polly
A dog par excellence!

All
He should keep his vows
Is it so much to ask?
He acts like a louse
We should bring him to task!

Lucy I wonder how many others there are, dropped across town like old Travelcards?

Polly More than a few, by the look of it.

Mrs Peachum (*sighs*) More than a few.

Filch enters with a tray of drinks.

Polly Ta, Filch. Let's get pissed.

Mrs Peachum Filch isn't like the others. You're a good lad, ain't you?

Filch I try me best.

Lucy Ooh, I'd like to get me hands on him!

Filch Who?

Mrs Peachum The Captain. – But no one know where he is!

Filch The word is, he's meeting the Gang.

Polly I don't know what I'd do if I saw him. I don't know if I'd laugh or cry.

Filch Polly . . .

Filch draws Polly aside. Lucy and Mrs Peachum hit the drinks.

Filch HOW ABOUT ME?
    He was my competition
    He was my bête noire
    Now he's out of the frame

Polly
    What's your game?

Filch
    You were quite captivated
    It was perfectly clear
    Now you're suddenly free

Polly
　Oh, I see . . .

Filch
　I've adored you for ever
　Especially your legs
　And I'm being sincere

Polly
　Well that's clear

Filch
　Would you, Polly, consider
　That going with me
　Might be awfully nice?

Polly
　Filch – no dice

Filch Well, look, think about it, will you?

Polly (*smiles*) You don't pick me up on the rebound, all right? I ain't a rebound kind of girl.

Filch I'd do anything for you. You know that.

Polly Anything?

Filch Yeah.

Polly Like go to the meet with the gang, and tell us where Macheath is?

Filch What, so you can get back together with him?

Polly Who knows? Want to take a chance? You might worm your way into my affections.

Filch I ain't a grass, Polly.

Polly Mum, is Filch a grass?

Mrs Peachum Yes.

Polly (*to Filch*) Then if you want to serve me, you know what you must do . . .

Polly gives him a peck on the cheek, and rejoins the others.

Filch I'm a fool for love. That's what I am.

SCENE TWELVE

Near North Greenwich Tube Station, the Dome looming in the background. Middle of the night. Macheath enters, at a run, shivering in his shirtsleeves.

Macheath I never knew England could be so cold. What happened to global bloody warming? I'm going to have to find some shelter, or I'm going to die a loser. Which would be a sorry conclusion, wouldn't it, to an uplifting tale? – I had to keep off the highway, for obvious reasons. I hacked along by the river, down Galleons' Reach, past the Barrier, till I come into the shadow of seven hundred and fifty million quid's worth of, er . . . tent. You can see it from outer space, apparently. Them Martians'll think it's the seat of world government. When in fact it's a hamburger stall. Seven hundred and fifty million quid? All I can say is, someone's doing very nicely thank you. And it ain't little kiddies in hospital, is it? Blimey, I wish I'd been the middle-man on *that* one. Ah, there's a bit of warmth from somewhere. Let's see if we can doss down here for a bit . . . and figure out what to do next . . .

He heads for some cover at the side of the building. He stumbles over someone in the dark.

Beggar 1 Watch it, pal!

Macheath leaps away in surprise, only to tread on someone else lying on the ground.

Beggar 2 What's your game? You're like a herd of elephants!

Beggar 3 Fuck off out of it!

Macheath Sorry, mate.

The Beggars – for we shall call them Beggars – are now slowly revealed. A large group of homeless people, old and young, lie in filthy sleeping-bags, up against the walls of the gleaming new architecture.

Blimey.

Beggar 2 Look, dumbo, these pitches is taken! So sod off!

Macheath (*bristles*) Do you know who you're talking to?

Beggar 2 Why, who am I talking to?

Macheath realises how comprehensively unarmed he is.

Macheath Don't matter. Nobody. I'll shift.

A NAUGHTY WORLD

Car keys – none
Side-arm – none
Mobile – none
Wallet – none
Steel comb (sharpened)
Signet rings
Pepper spray
Personal things –
Stripped of me accoutrements
Naked in a naughty world

He moves away. An old tramp – Dora – hisses at him.

Dora Psst! Over here, son.

Macheath What?

Dora You're going to get hypothermia.

Macheath Yeah, I know.

Dora You're going to die of exposure. Come. Come!

Macheath sits next to her.

Take this blanket. It's a good 'un.

Macheath What about you?

Dora I'm fine. I've got the *Financial Times* inside my bra.

Macheath Does that help?

Dora It's a quality paper. (*She hands him the blanket.*) Wrap it round you, go on.

Macheath By God, it ain't been washed for a while.

Dora I didn't say it was freshly laundered. I said it was warm. That better?

Macheath It is better. Thanks.

Dora Name's Dora. Pleased to meet you.

Macheath Macheath.

They shake hands.

How much you want? For the blanket?

Dora You can have it.

Macheath You can't just give it to us, for nothing.

Dora shrugs.

You could charge more or less what you like, Dora, and get away with it.

Dora Well, I know that. But I suppose I'm here because I don't really want to.

Macheath Too right; if that's how you think, it ain't surprising you've ended up dossing.

Dora What about you?

Macheath Me?

Dora Why have you ended up dossing?

Macheath Well, I'm down on me luck, aren't I.

Dora (*chuckles*) Haven't even got a blanket.

Macheath Are you laughing at me?

Dora I am indeed.

Macheath Not many gets away with that.

Dora Give it me, then. (*She holds out her hand for the blanket.*)

Macheath . . . No, no, I'm just getting some feeling back in me limbs. You laugh away, have a right good giggle.

Dora Perhaps that's the price you have to pay, Macheath.

Macheath . . . I was all set for the biggest deal of me life, in about – oh – five hours' time. All set, I was. All sorted. Getting into bed with the big boys. And then fate dealt me a cruel blow –

Dora No? Truly?

Macheath – yeah, and I was betrayed by a cunning little Vixen. And the whole edifice come crashing down round me ears. And now I'm here all cold and snivelly and grateful for a filthy old rug. I feel sorry for meself. I do.

Dora hands him a disgusting bottle.

Dora Try this. I recommend it.

Macheath Earlier on I was drinking champagne, at seventy-five quid a throw.

Dora That was earlier on.

Macheath . . . So it was. (*He drinks.*)

Dora It warms your guts, and gradually it dulls the memory, which I might tell you is a blessing. Feel better?

Macheath Yeah.

Dora Three men have died wrapped in that blanket. All of the people I give it to seem to swiftly shuffle off their mortal coil. What sort of life did you lead, then? Before you came down in the world?

Macheath I was a villain.

Dora Good at it?

Macheath Improving.

Dora Don't you want to know how they died?

Macheath No, not much.

Dora Impressive.

Macheath I ain't going to die.

Dora Never?

Macheath Nope. I'm charmed. – So what do you do, if we're having a conversation? Hang around Covent Garden with a dog on a string?

Dora I'm a beggar.

Macheath That's not a bad little number, begging. Who's your chief?

Dora I haven't got a chief. There is God. But he's tired of me.

Macheath Well, who do you report to, then?

Dora Eh?

Macheath (*as if to a child*) To who do you pay your commission?

Dora No, I'm a beggar. I beg. Like most of us here. We beg.

The Beggars WE HOLD OUT OUR HANDS
We hold out our hands
In hope that you'll see
Past the life that you have
To what might have been
To the end of ambition
The end of success
To pride's dissolution
Close knowledge of death

We hold out our hands
Not to hurt or to steal
But in honest cognition
That we have failed
And failure is cursèd
And failure despised
On the streets of New Britain
We are reviled

Dora
I am no angel
I knackered my life
I haven't a future
I'm not worth a light

The Beggars
We hold out our hands
In hope that you'll see

Past the life that you have
To what might have been
To the end of ambition
The end of success
To pride's dissolution
Close knowledge of death

We hold out our hands

Macheath Are you telling me that beggars is simply that? Beggars?

Dora I can only speak for those I know.

Macheath I thought it was the mother of all rackets!
I work with this bloke, he hustles Old Compton Street
in a mac tied up with a dressing-gown cord, back home
he's got a Beamer in the drive.

Dora I can only speak for those I know.

Macheath Why don't you all get together, all band
together like, and instead of asking for it, walk right up
and take it?

Dora What, and break the law?

Macheath Christ. Now I've heard everything. What the
fuck have you got to lose?

Dora Wrap yourself up warm and have a good old
drink. This is what life is. You'll get accustomed to it in
a year or two.

Macheath No way, mate. No way I'll get accustomed.
(*He stands up.*) I'm a doer, I am. I'm a fucking social
climber and proud of it. I'm –

*Across from them, there is an altercation. A Drunken
Beggar is pawing at a girl, and a woman is trying
to fight him off. The girl is Georgina and the woman
is Anne.*

Drunk Come on, little one!

Anne Get away from us! Get away!

Drunk I fancy you! Give us a feel, go on!

Georgina Don't touch me!

Anne Get off, you bastard!

Drunk Look, what's your problem, lady?

Anne You can't have her! Go away, please!

Macheath goes across. He carries Dora's bottle.

Macheath (*to the Drunk*) Back off, piss-pot.

Drunk You what?

Macheath You heard.

Drunk You starting something?

Macheath (*raising the bottle*) Yeah. Want to try me?

Drunk . . . Ah, farts to the lot of youse. Farts to the lot of youse!

The Drunk moves away and lies down, mumbling.

Anne Thank you so much.

Macheath Never could resist a damsel in distress. What's your name?

Georgina Georgina.

Anne She's my daughter, so . . . (*Pause.*) I know you, don't I?

Macheath Do you?

Anne You were in my garden a few hours ago. A lifetime ago.

Macheath (*He remembers. Grins*) Oh yeah.

Anne And now we're here.

Macheath I know why I'm here. I got a rough idea why they're here. But why the bleeding hell are you here?

Anne Georgina ran away.

Georgina Yes, and I'm not going back!

Anne . . . That's what she says.

Macheath Tell you what, Georgina, if I was quartered in a barracks like that, I think I'd put me feet up and stop complaining.

Anne (*sighs*) She won't be budged. So what can I do? All I can do is stay with her. And try and keep her from harm.

Macheath Out here in the cold and shit?

Anne Anywhere. What else can I do?

Macheath . . . Wish I'd had a mum like you.

Anne PARENTS AND CHILDREN
A mother's love is very hard to bear
A burden that brings with it woe and care

Georgina
A daughter's love is equally heartfelt
But I must get away to find myself

Anne
You're in danger in this wilderness
Come home with me and we'll clear up the mess

Georgina
I never want to see the place again!
Can't live a life of lying and pretend!

Both
Parents and children
One thing's for sure
Parents and children
Will go to war

Georgina
> I love you but you lay down all these rules
> What I can eat and drink and wear to school

Anne
> I love you but you cause me so much pain
> Why can't you be my baby girl again?

Georgina
> I'm growing up, I'm learning to be me
> I may not turn out as you'd like me to be

Both
> Parents and children
> One thing's for sure
> Parents and children
> Will go to war

Anne You have to come to terms with reality!

Georgina Why?

Anne Because it is the way it is, Georgina! It is the way it is. I don't like it any more than you do.

Georgina You cover for him. You tell lies all the time. You know where it all comes from. We live on dirty money.

Macheath I wouldn't be too picky if I were you.

Georgina Why not? Can't I have principles?

Macheath Well, you can, but they'll get in the way when you get older.

Georgina I can't live a double life!

Anne We must set our sails according to the wind.

Georgina No! I won't! I want to be a good person, a decent person!

Macheath (*to Anne*) You sure she's at the right school?

Georgina I'm not living on the spoils of organised crime!

Anne That's enough.

Georgina I have to make up this fiction about why we're so rich. I lie to my friends, I lie to everybody! At least here I don't have to lie. The world you and Father inhabit is a disgusting sham!

Anne Be quiet.

Georgina It's so deceitful! Everything is a lie! What he does for a living, where he goes at night – all bullshit! And what he does in the garden!

Anne Georgina, I forbid you to say any more! We don't know this man from Adam!

Georgina I'll say what I like! I'll tell him about the roses!

Macheath Eh?

Anne Georgina! You must learn to keep your mouth shut.

Georgina Why? Why am I denied a voice?

Macheath Freedom of expression. All for it.

Georgina I'm seventeen!

Anne But he'll rob us!

Georgina Good!

Macheath Roses? You mentioned roses?

Georgina We'd be better off without it!

Macheath Tell me about the roses, sweetheart.

Georgina Even the gardening's a fraud! He doesn't enjoy it at all! He hates getting dirt under his fingernails!

Macheath So Charlie Dimmock he ain't. Well, well . . .

Georgina It's just a front. Down below the rose-bed is where he hides his stash.

Anne Oh, you stupid girl! I have never been able to trust you!

Georgina I'm not going to live like you.

Anne In that case, I'll leave you here to take what's coming! And I hope it teaches you a damned good lesson!

Georgina I don't care! I know who I am here!

Anne You know nothing, you spoilt little brat!

> *Macheath has wandered away, grinning. We lose Anne, Georgina and the Beggars.*

Macheath ONE BIG DEAL – REPRISE
    I ride the wheel
    Going down, going up
    And now I can steal
    All the wine from the cup!

    Just one big deal
    With no parallel
    Just one big deal
    And I'm out of this hell!

Chorus (*unseen*)
    Just one big deal
    And you bid farewell
    Just one big deal
    To the carousel!

    Just one big deal
    And they write your name
    Just one big deal
    In the hall of fame!

    Just one big deal!

Macheath is walking, still wrapped in his blanket.
The sound of traffic. Cutting through the darkness,
the beam of a high-powered torch picks him out.
He freezes.

Mardell Stay where you are.

Macheath (*shielding his eyes*) Who is it?

Mardell The law.

Mardell comes into the light.

I've been looking for you, Macheath. Didn't think you'd get very far. And if it was me, I'd've made for the river. But why are you now walking *back* towards Woolwich?

Macheath It's the people. So warm and loving. I do like a hokey-cokey round the old joanna. Can't seem to keep away.

Mardell Crap.

Macheath Just you, is it?

Mardell Just me.

Macheath On your tod.

Mardell Where are you trying to get to?

Macheath Kent.

Mardell (*approaching him*) Why?

Macheath Pilgrimage. Got to see Canterbury Cathedral before I die.

Mardell I've been there. It's fine, but it's not first division.

Macheath Got a gun?

Mardell You said something to me. It stuck in my mind.

Macheath (*smiles*) No gun.

Mardell I wondered if you were serious . . .

Macheath What colour knickers are you wearing?

Mardell (*caught off guard*) Sorry?

Macheath takes a swing at her. He knocks the torch from her hand. He drop-kicks her and she goes down. Macheath throws himself on top of her, but Mardell gets a foot up and effortlessly levers him off. Macheath goes flying. Mardell is on her feet quickly. Macheath rises groggily and they circle.

(*Grins.*) My idea of fun.

Macheath I ain't going back inside, copper.

Macheath attacks. He punches, she ducks. Deftly, Mardell jabs at him with a series of well-time martial-arts moves. He's winded. She hefts him over her shoulder with a judo throw. He crashes to the ground. Mardell gets on top of him, and holds him in an arm-lock.

(*breathlessly*) Where the fuck did you learn that?

Mardell Hendon Police College. Now. I can arrest you. Or you can tell me if you were serious.

Macheath About?

Mardell About taking me on to your team.

Macheath (*amazed*) Do what?

Mardell I'm going to release you. But on your own, you're nothing. You haven't even got wheels. You need a friend, Macheath.

Slowly, she releases him. They sit looking at each other, panting.

Macheath You're one for the funny farm, you are . . .

Mardell (*laughs*) That's what my mum says. From the day I joined the force, she's maintained I was a lunatic. And now I think she's right.

Macheath Why?

Mardell Because it's all crap. What you've always known, I've just discovered tonight. It's crap. It stinks. There is no law and order. There's just a bunch of people on the make.

Macheath So . . .?

Mardell So I give up. To hell with it. I'm wasting my time. One day I'll die, and what will have altered?

Macheath Depends how much money you've got, don't it? Money alters everything.

Mardell THE TEMPTER
    The tempter comes
    In his hands forbidden riches
    The tempter comes
    He leads me to the edge

    The tempter comes
    And I look out on the kingdom
    The tempter comes
    It's just a step into the air

Macheath
    Well I believe that I can fly
    It isn't difficult, come take my hand and try
    There is no doom, no death, no divine punishment
    The promised land awaits for those who don't repent!

Mardell
    The promised land
    Can I forget what I believed in?

The promised land
I want to fly but I'm not sure!

Macheath
Well I believe in nothing more
Than in myself, and me, and how the secret door
May be unlocked, and with the villain's magic key
We enter realms of unimagined luxury!

Both
The tempter comes
You realise that you're gleeful!
The tempter comes
There is no good and evil!
Throw off your tattered cloak of moral laws
Your urge to be a winner is primeval
So bare your teeth, start sharpening your claws!

Macheath You want to join my gang.

Mardell Did I say anything about a gang? I just want to join you.

Macheath . . . I'm not a very reliable sort of bloke, you know.

Mardell But you've got something up your sleeve, haven't you? Something's going off down in Kent?

Macheath Yeah. Something big.

Mardell Want to cut me in?

Macheath What, just like that? I mean, why should I?

Mardell Well, you fancy me, for one.

Macheath I fancy anything with tits – don't mean I do business with them. Customarily my blokes are blokes.

Mardell (*dangles car keys*) It's a good few miles, Kent.

Macheath You might drive me straight down the nick.

Mardell You can drive. You're the boss. (*She throws him the keys.*)

Macheath Was this on the course at Hendon, too? The Road to Damascus routine?

Mardell Might have been. Might be going under cover, mightn't I? Might all be a trap.

Macheath Yeah, so how do I know if I can trust you?

Mardell You don't. You know what? I think I'm being profoundly stupid. Forget it. We never met.

She walks away.

Macheath You ain't half got a gorgeous bum.

Mardell laughs, and turns.

That's a real villain's bum, that is.

Mardell I can't just throw away my life, can I? How can I do that?

Macheath Easy.

Mardell Maybe I'm on a mission. To kill you.

Macheath Yeah, I thought of that. But you would've done it by now.

Mardell . . . I fancy you.

Macheath . . . Now we're getting to the heart of the matter. Come here.

Mardell Why?

Macheath I need to check you ain't wearing a wire.

She goes to him. Macheath runs his hands over her body. It turns into an embrace. They kiss.

Christ. You are a serious kisser.

Mardell This is a mistake, isn't it?

Macheath It's a terrible mistake.

Mardell It's against all the rules.

Macheath That settles it. Let's go.

They exit, taking the torch. The sound of a car starting up.

Chorus (*sweet and low*)

ONE BIG DEAL – REPRISE

Just one big deal
Risk all you've got
Just one big deal
And walk off with the pot!

Just one big deal
Just one big deal . . .

SCENE THIRTEEN

Macheath and Mardell are in an unmarked police car on the motorway. Macheath drives. Mardell smokes. Occasionally, the headlights of passing cars wash over them.

Mardell It spreads from the top down, the corruption. Politics, business, the force. The bloody opera for all I know. Every time you open a paper it's sitting there in front of you.

Macheath What paper do you read? My one's mainly football and the Royals.

Mardell They say you were in the Army.

Macheath Yeah, I was, yeah.

Mardell With the rank of Captain.

Macheath Well . . . almost.

Mardell So what turned you to crime?

Macheath What is this, *Newsnight*? Nothing turned me to crime. It's in me bleeding DNA. I'm a villain, born and bred.

Mardell It must be nice, to know where you belong.

Macheath Yeah. Sweet.

Mardell Ever since I left college I've been trying to find out. Trying to fit in. Assimilate. I've been as pure as the driven snow. Virgin white.

Macheath Virgin black.

Mardell (*grins*) Hardly.

Macheath You saucy little devil.

Mardell Anyway, I failed. I lost a pointless battle. I can't make the slightest difference to the systems of hypocrisy, the universal fiddle. So I'll be as bad as all the rest. No. Badder than the rest. What's your Christian name?

Macheath What's yours?

Mardell Guess.

Macheath Euphemia.

Mardell No.

Macheath Jocelyn.

Mardell (*laughs*) No.

Macheath Ethelberta.

Mardell No!

Macheath Anastasia.

Mardell Nearly.

Macheath Mary Poppins.

Mardell Not likely!

Macheath Give up.

Mardell Stacey.

Macheath Nice. Rhymes with tasty.

Mardell No it doesn't. What's yours, then?

Macheath Me what?

Mardell Your name.

Macheath I haven't got one. I'm the Captain.

Mardell I can't call you that.

Macheath Everyone calls me that. It's all about respect, see. You don't want to give away too much information. Too many personal details. You don't want to have no credit cards, no identity cards, no saver cards at the supermarket. You don't want them to know what you're buying, what you're spending, what you're thinking about in the bath. You don't want to be in no one's computer. If you can possibly help it. You don't want a fixed address or a kosher driving licence. You don't want tax returns or mortgages or stupid fucking air miles. There's only one way to live in civvy street, and that's as a free man, what nobody ain't got tabs on. With as many aliases as the death columns in the *South London Press* can supply you with. And all your money in cash. I have never found my role in society, as several probation officers have said. I looked for it, but I can't find it. All I can find is duckers and divers, predators out for themselves. And I reckon the best way to keep one step ahead of Big Brother is to never let nobody know who you are. Not even in action. Not even in bed. That way, you get respect. But what happens is, you end up not knowing who you are yourself. You've used one alias too many. You've lost your bleeding compass, you don't know up from down. So you stagger from this job to that, getting a result here, a result there, a stretch if you're grassed – going round and round like a fucking hamster in a fucking cage. And I want out. I want one big score, and out. When I'm out, I'll tell you me Christian name, Stacey. If you're still interested. Here, you mind taking your hand off me leg while I'm driving?

SCENE FOURTEEN

Mr Big's garden. Moonlight. Mr Big is in his silk pyjamas and dressing-gown, a drink in his hand, gazing at the stars.

Mr Big STARS
  I don't know their names
  At school I never listened
  The enigmas of the cosmos
  Passed me by
  One of them's the Plough,
  One of them's Orion
  But in my ignorance
  They are just stars to me

  But when you see them
  In their millions
  When you count
  And give up counting
  When you picture
  Civilisations
  With their heroes and their villains
  And their unknown histories
  You know exactly how it feels
  To be alone

You know exactly how it feels
To be alone

He hears a noise, off. The clunk of metal on metal.
He frowns.

(*Calls, softly.*) Anne? (*Pause.*) Georgina?

He frowns again, and goes quickly into the house.
    Macheath enters from the garden, furtively. He's
muddy. He carries a garden fork, and a large metal
suitcase.
    Mr Big comes out of the house with a shotgun.
He sees Macheath and levels the gun at him.

Mr Big Nice try, pond-life.

Macheath Oh, shit.

Mr Big How did you know it was in the roses?

Macheath Met your daughter as a matter of fact.

Mr Big Put it down.

Macheath puts the case down.

And the fork.

Macheath puts the fork down.

Lie on the ground. On your face! Taste the soil of Kent,
Macheath.

Macheath lies down. Mr Big puts the shotgun to his
head.

Macheath Don't kill me, Mr Big.

Mr Big Of course I'm going to kill you. I've got to kill
you, haven't I? *Pour encourager les autres.* We go to
France quite often. It's two hours on Eurostar from
Ashford. Smashing. And you have to try, don't you?
With the language?

Macheath Yeah.

Mr Big Do you have languages?

Macheath I can do Estuary-speak.

Mr Big (*laughs*) You're a bright little bastard. It's a shame you've overstepped the mark. It is such a shame, when my employees double-cross me, and I have to prematurely retire them. But there you go. One of the indignities of management. Get up and sit in that chair. Now!

Macheath does so.

Where are my wife and daughter?

Macheath Up Greenwich, by the Dome.

Mr Big What the fuck are they doing there?

Macheath Sleeping in cardboard boxes.

Mr Big Christ, I hope no one finds out in the village.

Macheath I won't tell a soul, honest.

Mr Big Too fucking right you won't. Where exactly are they?

Macheath I'll take you.

Mr Big Sit down! I will not be double-crossed!

Macheath With respect, sir, you blanked me this evening. You turned me over to the filth! And all you had to do was back my alibi!

Mr Big What, and be interviewed, give statements, go to court? Are you out of your mind? I'm judging the gymkhana next week!

Macheath Why do you think Georgina hates you so much?

Mr Big I don't know. Fucking hormones or something.

Macheath No. It's 'cause you're a villain. Like me. Only I ain't trying to fit in. I ain't got no contradictions. Therefore, I ain't got no weak spot. Therefore, you should take me back, and give me a second chance.

Mr Big (*laughs*) Weak spot? You do have a weak spot, Macheath. Right here. (*He puts the shotgun barrel to Macheath's head.*)

Macheath She's a bit tasty, your wife. I can see why you dote on her. Definite class.

Mr Big I know. Nobody'd give me the time of day if I wasn't with Anne.

Macheath (*sighs*) No gymkhanas . . .

Mr Big No.

Macheath No harvest festivals . . .

Mr Big No.

Macheath There is something about a real blue-blooded tart, isn't there? Makes you feel you belong? Why don't you let me show you where she is?

Mr Big Because I don't crawl to women. If they want to come back, they can come back. Do you know how I got to where I am?

Macheath I think I can guess.

Mr Big By being totally, utterly ruthless. (*Points the gun.*) On your knees.

Macheath kneels.

Say your prayers.

Macheath Can't remember any.

Mr Big No, nor can't I. You know what? I'm quite envious of you, Macheath.

The gangster knows his end is coming
From the day that he is born
It will be swift, and cruel, and sudden
There won't be many who will mourn

Macheath
The gangster knows that death is ready
To take him at any time
And from the moment that he's dead he
Is absolved of every crime

Mr Big
The gangster meets his God with laughter
Pressure's off, he can relax
And he'll stay happy ever after
Secure from knife, and gun, and axe . . .

Both
It's fucking great in gangster heaven
Big-shots sit around and chat
Cocktails start at half-past seven
No one fears they will get whacked!

Mr Big Time for you to go there, pond-life.

*Mr Big cocks the shotgun and aims it. Mardell enters
and shines her torch directly into Mr Big's face. He's
blinded.*

Mardell Hey!

Mr Big What the fuck – ?

Mardell Run!

*But Macheath doesn't run. He grabs the garden fork,
swings it, and clatters Mr Big to the ground. The
shotgun goes off wildly. Macheath gets to his feet with
the fork. Mr Big groans.*

Come on! Run!

Macheath stands astride Mr Big with the fork in his hands.

Run!

Macheath No one calls me pond-life.

Macheath raises the fork with both hands, and as he plunges it down into Mr Big's chest we blackout. Mardell screams.

SCENE FIFTEEN

The underground car park. The Gang come out of the shadows. Filch is with them.

Gordon Did the Captain say what he wanted?

Steve No. He just called a meet. But there's something going off, I can smell it.

Ryan Macheath's a master of intrigue. He's give the bill the run-around again!

Jack And Peachum too!

Laetitia What an artist!

Pinhead enters.

Harry Wotcher, Pinhead.

Pinhead We back in business?

Steve Yeah, the Captain's on his way. What happened to you?

Pinhead I got sussed, Steve, down the Trafalgar Tavern. I got banged up, and the Captain, God bless him, helped me escape. What's the plan, then?

Laetitia We dunno.

The sound of a car drawing up nearby.

Steve Looks like we'll soon find out.

Macheath enters, carrying the metal suitcase. Blood on his white shirt.

Good to see you back, sir!

Pinhead Three cheers for the Captain!

They all clap Macheath on the back. He's pleased.

Macheath I have returned to resume command. We are going on a big operation. We are going behind enemy lines.

Steve When, sir?

Macheath Immediately.

Macheath opens up the suitcase.

Now, in here is about three million quid in used notes.

Susie Whose is it?

Macheath Mine. We're moving up into the big league, ladies and gentlemen. And I'm taking over the Gang.

General concern.

Harry What about Peachum?

Macheath Peachum's dead in the water. Look: I have his files. (*He takes the organiser from inside the suitcase.*)

Harry He didn't look dead in the water two hours ago. Looked in the pink to be honest, didn't he, Filch? I ain't sure I can just walk out.

Macheath (*consults the organiser*) Harry Paddington. Here we are. Well short of half-yearly targets. Will not get productivity bonus. To be sentenced for aggravated burglary at next Quarter Sessions.

Harry No! After all my years of service!

Pinhead Would you Adam and Eve it?

Gordon It's nae possible!

Macheath (*consults the organiser*) Scotch Gordon. Victim of fluctuating market forces. Viability: nil. Awaiting vacancy at Barlinnie.

Gordon Not back across the border! Anything but that! It's so frigging cold!

Macheath Your liberty hangs by the flimsiest thread. Every single one of you. Peachum will shop us all if it suits him. I am offering you the chance to make a cool million. Each.

Perry A cool million? Each?

The Gang A COOL MILLION
  A cool million
  All we've ever wanted is
  A cool million
  All we sweat and slave for is
  A cool million
  It's what we wish for when we blow the candles out
  It's what we whisper when we sit on Santa's knee
  It's what we toss and turn in bed and dream about
  A cool million quid in unmarked currency!

Macheath It will be risky. Riskier than the usual, 'cause we ain't dealing with amateurs here. But I will reward you handsomely. So – are you in?

Jack I dunno, Captain. How are we going to make this dosh?

Macheath There's a plane landing at dawn. Essex. We're going to score more coke than a rock band on a world fucking tour! By the time we've cut it and sold it, I reckon we quadruple our stake. Minimum!

Jack But Mr Peachum always done all right by me. He sends the wife a box of crackers every Christmas.

Laetitia Come off it, Jack. You nick a three-hundred-pound car stereo, he gives you thirty quid! It ain't exactly a fortune, is it?

Perry I like knocking out people's fireplaces. It's a laugh. Don't know if I'm ready for the big stuff.

Steve I reckon I am.

Susie Me too.

Filch Peachum's a twat. I vote for the Captain.

Perry I dunno, though . . .

Jack If it's risky . . .

Macheath All right. Now I'm going to tell you something you might not want to hear. (*He holds up the organiser.*) I need back-up. I need a crew I can trust. Either you come with me, or . . . I may have to resort to a little bit of grassing of me own.

Steve Well, that's made things perfectly clear.

Gordon We'll take the million!

Laughter. Macheath joins in.

Macheath Knew I could count on you . . .

Steve Course you can, guv!

Macheath takes pistols from the suitcase, and hands them round.

Macheath I don't know exactly what to expect, so we better go prepared.

Macheath offers Pinhead a pistol. He refuses.

Pinhead I never touch them, Captain, sorry. It's my religion, see. I'm in this weird sect that wears mauve on Wednesdays and –

Macheath Well, if you get shot, you get shot. (*He moves on.*)

Pinhead Captain?

Macheath Yes?

Pinhead How long are we going to be in Essex? I mean, what part exactly have we got to get to? Only I promised me auntie I'd take her up Smithfield for a pound of mince – she's got a bad leg, see – and they start early as you know, so I'm like wondering if –

Macheath Fuck Smithfield, and fuck your auntie.

Pinhead I don't think you'd want to, sir.

Macheath You can buy her a butcher's shop if this comes off. Steve, we'll go in three cars.

Steve You coming with me?

Macheath No. I'm sorted. We're looking for a field east of Woodham Ferrers, it's pasture, it's a mile along the B-road from a pub called 'The Three Jolly Brokers', and there's beech trees either side, if anyone knows what a beech tree is. So we'll rendezvous outside this boozer at 0500 hours. Any questions?

Susie We got time for a line before we go?

Macheath No. You can get caned after. Don't look for me, I'll be out of the country. Now check them shooters, and watch your backs. – Oh, and I want to introduce you to someone.

Macheath whistles. Mardell enters, cradling Mr Big's shotgun in her arm.

This is my new associate. Stacey Mardell.

Mardell Hi.

Jack She's a copper.

Laetitia I've been strip-searched by her.

Macheath She's in the squad.

Consternation.

Steve This ain't on!

Gordon No way!

Mardell throws the shotgun to Macheath. He catches it deftly and cocks the hammers, surveying the Gang. They go quiet.

Macheath She saved my life. She done stuff no copper would ever do. So she's in the squad. All right?

There's no more opposition.

All right. Move out!

The Gang exit quickly. Macheath closes the suitcase. Mardell comes to him.

Come here, baby.

They kiss. Filch, lingering, sees this as he leaves.

SCENE SIXTEEN

*The club in Mayfair. Roy and Neil still sit at their table,
completely drunk, and only just conscious. Vixen poses
on the table. The club is otherwise empty apart from
Miss Cinnamon drinking a mug of cocoa and reading a
Jilly Cooper novel, and Angelo and Freak, who play
cards and smoke. The Bouncer lounges by the door.*

Vixen
    'The raven himself is hoarse
    That croaks the fatal entrance of Duncan
    Under my battlements. Come, you spirits
    That tend on mortal thoughts, unsex me here
    And fill me from the crown to the toe top-full
    Of direst cruelty. Make – '

    Warren enters at a run.

Warren Murder!

Cinnamon (*alarmed*) What?

Warren The boss has been murdered!

    *Vixen gets off the table and joins Miss Cinnamon.
    Warren goes to Angelo and Freak.*

Angelo That's impossible!

Warren I swear to God, he's dead!

Freak Who done it?

Warren That little jam-rag what was in here earlier.

Freak Macheath? But he was arrested!

Warren He must've got out!

Angelo How do you know? Was you there?

Warren No, I was kipping, but I hears a shotgun blast! I finds him in the garden, with three prongs through his heart! As I cradles him in my arms, he says: 'Macheath done this.' And then he dies!

Angelo What's going on, Warren? We was all hunky-dory this morning.

Warren I don't know, mate.

Angelo Something's happened to the natural order of things.

Warren It's unforgivable!

Angelo It's like the lights has gone out all over the city. All the traffic signals, all the neon signs. And all the shop windows. Dark. It's like a blanket of night has descended.

Warren As I was driving up the Old Kent Road, three foxes ran in front of me motor. Three! What do you make of that?

Freak Spooky, man.

Angelo We can't tolerate this. We must have vengeance.

Warren Yeah. Or Christ knows what'll happen next. Armageddon and whatnot.

Angelo If he wants a war, he's got a war.

Warren He took the boss's money. I know where he'll be headed.

Angelo What car you got?

Warren Range Rover. On the double yellows.

Angelo OK. Tooled up?

Warren draws a sawn-off shotgun.

Freak?

Freak draws an Uzi. Angelo also draws an Uzi.

Freak You can't kill the king. It isn't done. We'll stick his fucking head on a pole.

Warren And shove his heart up his arse.

Angelo Let's not hang about, then.

As they turn to leave, Cinnamon approaches them.

Cinnamon Boys, take me with you.

Angelo Miss Cinnamon?

Cinnamon I too want the blood of Macheath. Your boss was dear to me.

Warren Was you . . .?

Cinnamon For fifteen years.

Warren Would you credit it?

Cinnamon He was my life. And now he's gone. And I'll have my revenge.

Angelo Got a weapon?

Cinnamon (*to the Bouncer*) Bobby. The gun.

The Bouncer produces an enormous assault rifle, and hands it to Cinnamon.

Angelo FIRE POWER
When it comes to armaments, we ain't exactly choosy

Freak
But if you've got an ounce of sense, don't argue with an Uzi!

Warren
A shotgun with the barrel cut fits nicely down your trousers

Cinnamon
I fancy a Kalashnikov will out-shoot Colts and Mausers!

Angelo

London has moved on a bit since Dixon of Dock
Green

Freak

We've got more fucking firepower than Morse has
ever seen!

Warren

From Hammersmith to Leytonstone, the air is thick
with shooting

Cinnamon

There's bullets flying everywhere, from Tufnell Park
to Tooting!

All

When we're going out to work
We definitely figure
Survival is more likely with
A finger on the trigger!

Cinnamon

Well, you love it on the telly when John Wayne
shows up mob-handed

Freak

Now it's coming home to roost; the fucking eagle's
landed

Angelo

America has changed us from the innocents we was

Warren

Into psycho gunslingers with dum-dum brains
because

All

When we've got a job to do
We definitely reckon

You get a better service with
An automatic weapon!

They check their magazines and flick their safety-catches off.

Angelo Let's go to work.

They exit. Roy blearily waves another £20 note at Vixen. With a sigh, she climbs back on their table.

Vixen
'Make thick my blood;
Stop up the access and passage to – '

Blackout.

SCENE SEVENTEEN

Outside 'The Flower of Kent'. Police sirens approaching. Then Lockit enters from the street at the same time as Peachum enters from the pub. They both wear raincoats, and are in a hurry.

Peachum I got the call! Woodham Ferrers, in Essex! You got back-up?

Lockit The Met's finest. They make Dirty Harry look gay.

Peachum He is, isn't he? – Come on.

They exit. Filch, Mrs Peachum, Polly and Lucy enter from the pub, pulling on coats. The Women are completely pissed.

Polly Where's he gone?

Filch Essex!

Polly Essex?

Lucy I never goes to Essex. They can't talk proper in Essex.

Filch He's meeting a plane!

Mrs Peachum What for?

Filch He's doing a drugs deal! And then he's going to leave the country with several million quid! And he's got a new girlfriend and all!

Polly What?

Lucy Sure it's bad enough leaving us in the lurch –

Polly – but if he thinks he's bunking off abroad with a fortune –

Lucy – and a tart –

Polly – he can bleeding well think again! Come on, let's go! Mum, give us your car keys!

Mrs Peachum (*hands over keys*) Can we have a sing-song while we're driving?

Filch Polly, you can't drive, the state you're in!

Polly Filch, do you love me?

Filch (*blushing*) You know I do.

Polly Right then, you drive!

Mrs Peachum How about 'E Viva Espana'? That'll bring back some memories . . .

> *They exit.*

SCENE EIGHTEEN

A field near Woodham Ferrers. Trees on either side.
Dawn is breaking. Birdsong. Then the sound of a light
plane taking off and flying away. Angelo's Mob enter at
one side, sneaking through the trees.

Angelo Make use of this cover. He's got to come across
the field.

Freak When he gets into view, we'll blast him!

Peachum enters at the other side with Lockit and two
heavily armoured Police Sharpshooters with high-
calibre rifles.

Lockit Take your positions. (*He draws a pistol.*)

Peachum Got one for me, Dick?

Lockit Sergeant?

One of the Sharpshooters hands Peachum a pistol.

Peachum Let me give him the *coup de grâce*, will you?
After all, it's my little girl he's been humping.

Lockit Well, he's been humping mine too.

Peachum Yeah, but mine was the one he married.

Lockit Mine was the one he got pregnant!

Sergeant (*looking through night-vision binoculars*)
Target approaching, sir.

Warren Look there! Someone's coming!

Dim figures are emerging through the morning mist,
at the far side of the field. Everyone cocks their
weapons and takes aim.

Lockit When I give the word, fire one round to make him draw his weapon. I want him dead with a gun in his hand.

Sergeant Yes, sir.

Angelo He's got to know it was us, ain't he? He's got to know who put out the contract, and what a dreadful thing he done. Who's going to break cover?

Warren I will. I loved my boss.

> *Macheath and Mardell come out of the mist, struggling to move fast whilst carrying two large, heavy canvas bags each.*

Lockit On my order. Ready?

Angelo Get set, Warren.

Macheath There's the car!

Lockit Now.

> *At the same moment, a Sharpshooter fires a shot into the air, Warren breaks cover and levels his sawn-off shotgun, and Polly and Lucy run on and go straight to Macheath, followed at a distance by Mrs Peachum and Filch.*

Sergeant Armed police!

Warren You're dead, Macheath!

Macheath Jesus!

Angelo Who are them tarts?

Lockit Hold your fire!

> *Lucy and Polly lay about Macheath with their fists. No one can get a clear shot at him. Mardell throws herself to the ground. Warren ducks back under cover.*

Mardell Shooters to left and right!

158

Steve Captain! It's a trap!

The Gang emerge dimly through the mist to the rear.
They're all armed. (Pinhead isn't with them.) We
freeze the action.

Polly HE HAS TO DIE
   We are friends now
   Me and Lucy
   And we want you to believe us

Lucy
   That we'll never
   Trust a lover
   Who seduces and then leaves us

Polly
   Broken hearts are
   But a small thing
   When compared to how you've hurt us

Lucy
   We have sworn
   To never love
   A man so vile as to desert us

Lucy and Polly
   Don't come knocking round our door –
   We don't want you any more!

Macheath
   Lucy, my darling
   Polly, my angel
   I loved you with such fervour!
   Can I help it if my heart
   Is stolen by another?

Lucy and Polly
   Who?
   We'll break her arms
   And rip her lungs out

Macheath (*indicates Mardell*)
Her –
She is the one
For whom my tongue's out
Ain't she delectable?
Stacey Mardell . . .

Angelo's Mob
He deserves to die

Peachum's Mob
He deserves to die

Filch and Mrs Peachum
He's such a bastard!

Mardell
I know that he's a wicked, wicked man
But I can't help myself –
I love him!

Peachum's Mob
Oh you poor fool!

Angelo's Mob
Oh you poor fool!

The Gang
We understand –
He has this effect; there isn't a
Girl who ain't succumbed
To his charisma

Macheath
My loyal gang!
But it's too late
They intend to
Wipe me out

Angelo's Mob
We've no hesitation
He has to die

Peachum's Mob
>There's no obfuscation
>He has to die

Macheath
>You see?
>I have to die

Filch and Mrs Peachum
>He has to die
>His time has run out

The Gang
>He has to die
>His battle is lost

Mardell, Lucy and Polly
>He has to die
>His blood will soon spout

Angelo's Mob
>He has to die
>For killing our boss

Peachum's Mob
>He has to die
>And we don't give a toss!

All (*except Macheath*)
>He has to die
>He is scum
>He has to die
>The time's come
>He has to die!

Macheath
>I love my life!
>How I love my life!
>And I will never
>Never ever
>Say goodbye!

Peachum Polly! Get down!

Lockit Lucy! Get down!

Polly and Lucy hit the deck. Macheath is left standing alone. He draws his pistol and holds up his hands in a gesture of supplication. At that moment the firing starts. The gunshots are done on a soundtrack; the movements are balletic, almost slow-motion. There is a chaos of unbelievable firepower, smoke, darkness, shouts, screams. Everyone shoots at each other.

Polly and Lucy manage to crawl to Mrs Peachum and safety. Filch holds Polly. Police and Gangsters run out into the open, firing wildly at each other. Mardell is last seen crawling away on her belly, dragging a canvas bag.

When the shooting stops and the smoke clears, there are bodies everywhere, but no sign of Macheath. All of the Gang, both the Sharpshooters, Peachum, and Lockit, and all of Angelo's Mob have been hit.

The women are weeping. Peachum, grievously wounded, crawls through the carnage to the dying Lockit.

Peachum Did we get him? Dick?

Lockit Why is a cross-fire always withering? (*He dies.*)

Peachum Dick! I don't believe it! He can't have got away! (*He dies.*)

Macheath materialises on the forestage, picked out in a spotlight, separated from the others, who remain onstage behind him.

Macheath But I have.

CURTAINS

Though curtains looked inexorable
I seem to be invulnerable!

> Though logically I can't escape unharmed
> You cannot kill a villain who is charmed!

Pinhead appears behind Macheath and holds a pistol to his head.

Pinhead Yeah, I can, mate. Sorry.

Macheath Pinhead . . .?

Pinhead Yeah. It's a game, innit?

Macheath You sure it's loaded?

Pinhead Hundred per cent.

Macheath So this is the end?

Pinhead This is it. Sorry.

Pinhead cocks his pistol.

Macheath Oh, shit.

Just as Pinhead is about to fire, Mr Big descends from the flies, on a wire. He has sprouted little angel's wings. A spot on him; and the rest in darkness.

Mr Big
> One moment please! Although one of our basic
>   premises
> Is that at the finale, the crook meets his nemesis
> We don't want you lot leaving believing
> No villain makes a go of thieving –
> Well it wouldn't be true! And truth is of the essence!
> Despite that we've shown him to be an excrescence
> Macheath survives. The gun jams. Whatever.
> The assassin blows it altogether –
> And it's up to me, Mr Big, to fashion a
> Heart-warming end, like a *deus ex machina*
> Normally does. So join us once more, briefly,
> To witness the fact that reality chiefly

Turns out to be grossly unfair. Come and see where
Our lucky anti-hero's share has let him build his
   final lair!

*Tropical paradise music. Lights up to reveal Macheath
under a palm tree. He wears brightly coloured
swimming shorts and shades, and reclines on a
sunbed. Mardell enters, wearing a bikini. She carries
two enormous cocktails to Macheath. As she puts
down the drinks, he stands and takes her in his arms.
Macheath has his back to us.*

Macheath (*aside, over his shoulder*) Not bad, is it?

They kiss passionately.

Now that's what I call serious kissing.

Mardell laughs happily.

Not bad at all . . .

*They kiss again, Macheath keeping his back to us.
Mardell's hand comes up with a tiny silver pistol in it.
She holds the pistol to his head.*
   Blackout.

ENCORE

Mr Big leads the Company.

Company GANGSTER HEAVEN – REPRISE
   The Villains' Opera is over
   Everyone is on Cloud Nine
   The guy up there we call Godfather
   Never heard of closing time!

Mr Big
   So do you think we get to heaven?
   Do you think we are redeemed?

Do we deserve the angels' welcome?
Or are we the scum we seemed?

Company
We've really worked for our salvation
Now we're dead we are set free
It's fucking great, obliteration
No one's got no enemy!

Mr Big
There ain't a winner or a loser
Only villains with a thirst
We're off to the celestial boozer
Mine's vodka-tonic if you're first!

Warren Don't forget the lime!

Company
It's fucking great in gangster heaven
Big-shots sit about and chat
Cocktails start around eleven
No one fears they will get whacked!

Mr Big It's been a real pleasure, ladies and gentlemen.
Look out for the reviews in the *South London Press*!
And mind how you go in that car park! Goodnight!

The End.